Romanticism, Liberal Imperialism, and Technology in Early British India

"Deftly combining the history of technology with literary analysis, Daniel White's new book fascinatingly reveals the centrality of the steam engine to the British imperial imagination. Focusing on the speculative fiction of 1830s Bengal, White traces colonial visions of the future, with everything from air-conditioned trains to steam air balloons. A steam-powered tour de force of colonial literary history."

—Kate Teltscher, *University of Roehampton, UK*

"In this well-written, well-researched, and fascinating account, White offers steam as a way to rethink parallel literary and scientific histories that have had significant consequences for colonialism. In White's book, steam is an invention but also an idea, one that contributed to crucial debates about how technological development impacts liberal thought. With characteristically revealing detail, White gives readers a new vision of empire as a place for techno-futurism and its cautious appraisal, which contributes important lessons for our own age of buoyant invention. An unusual book, in the best way."

—James Mulholland, *North Carolina State University, USA*

"Daniel White's brilliant book explores futurist fictions published in *The Bengal Annual* and Romantic poetry side by side so as to dissolve the borders between metropolitan and colonial cultural production. Fascinating details emerge of an imperial imagination in which steam, in particular, was firmly fixed as a "romantic machine" that would diffuse European civilization, but which nevertheless generated unexpected perspectives on liberalism and the future of India. His enviable range of references across different liberal-imperialist visions of technology, global capitalism, and British-Indian nationalism informs and energises a discussion that also encompasses activities, friendships, and writings that are fun to read about even today."

—Rosinka Chaudhuri, *Centre for Studies in Social Sciences, Calcutta, India*

Daniel E. White

Romanticism, Liberal Imperialism, and Technology in Early British India

"The all-changing power of steam"

Daniel E. White
Department of English
University of Toronto
Toronto, ON, Canada

ISBN 978-3-031-60704-2   ISBN 978-3-031-60705-9   (eBook)
https://doi.org/10.1007/978-3-031-60705-9

Cover credit: © John Rawsterne/patternhead.com

This Palgrave Macmillan imprint is published by the registered company Springer Nature Switzerland AG
The registered company address is: Gewerbestrasse 11, 6330 Cham, Switzerland

If disposing of this product, please recycle the paper.

For Ginny, Camilla, and Nora

ACKNOWLEDGMENTS

While researching and writing this piece, I have incurred debts of gratitude to Rosinka Chaudhuri, Jeff Cox, Nora Crook, Lindsey Eckert, Joshua Ehrlich, Mary Ellis Gibson, Audrey Jaffe, Greg Kucich, Margaret Makepeace, Máire ní Fhlathúin, Michele Speitz, Nathan Wood, my graduate students in Fall 2023, and the staff members of the Rare Book Room at the British Library. I appreciate each of you.

CONTENTS

LIST OF FIGURES

CHAPTER 1

Introduction: Motions and Means

Abstract After briefly sketching the development of steam power in Europe, the Introduction depicts three remarkable scenes inaugurating steam power in British Bengal. I next offer an account of Romantic figures fascinated by steam. We then turn to Calcutta, where anglophone writers were relentlessly projecting utopian or dystopian futures powered by steam. *The Bengal Annual* for 1835 includes two such works, Henry Hurry Goodeve's "1980" and Henry Meredith Parker's "The Junction of the Oceans. [*A Tale of the Year 2074.*]." These works represent distinct perspectives within the prevalent liberal imperialism of the period. The Introduction closes by presenting the structure and main argument of the subsequent chapters. After focusing on a Romantic strain of thought according to which steam technology represents an anti-utilitarian humanization of nature, the Pivot will place within and against that frame the uneven and divergent ways in which writers in British India map a constellation of liberal imperialist values on to their hopes and fears concerning a future powered by steam.

Keywords Steam power (depictions in Britain and India) · Speculative fiction · Liberal imperialism

Back in England in 1837 after twenty-two years in Calcutta, a "returned exile" informs his metropolitan readership that "the all-changing power

1
D. E. White, *Romanticism, Liberal Imperialism, and Technology in
Early British India*, https://doi.org/10.1007/978-3-031-60705-9_1

of steam has performed its metamorphoses in India as well as in Europe" ("Reflections," 1837, pp. 65–66). "TO STEAM OR NOT TO STEAM, THAT IS THE QUESTION," proposes the writer of a letter to David Lester Richardson's *The Calcutta Magazine* in 1830, answering in the affirmative that "being in effect brought nearer to our native homes by many thousand miles, and being able to reach it in person or by letter in the brief period of six weeks instead of that number of months," constitute "a consummation so devoutly to be wished" ("Steam Navigation," 1830, p. 319). In both parts of the world and promising to abbreviate space and time between the two, steam performed its metamorphoses on the totality of early-nineteenth-century "Motions and Means" (Wordsworth, 2004, p. 604), the opening phrase of Wordsworth's sonnet, "Steamboats, Viaducts, and Railways" (1835), including the imagination. Steam engines, further, could even *have* an imagination. Readers of the popular *The English Spy* (1826), illustrated by George Cruikshank, encountered the following lines from Charles Molloy Westmacott's poem "The Advantages of Steam," at the end of which steam powers a story engine, prefiguring contemporary anxieties about artificial intelligence supplanting not just human productivity but even human creativity:

> Arts, manufactures, coaches, ships, alike impell'd by steam;
> Fire and water changing bubbles into gold.
> Steam's universal properties are every day improving,
> All you eat, or drink, or wear is done by steam;
> And shortly it will be applied to every thing that's moving,
> As an engine's now erecting to write novels by the ream. (Blackmantle, 1826, p. 143)

Considering metropolitan and colonial cultural production as a "unitary field of analysis" (Cohn, 1996, p. 4), this Pivot shows how tensions in the 1830s between utilitarian and Romantic perspectives on steam power marked meaningful divisions within the pervasive liberal imperialism of the era and generated divergent speculative fantasies about the future of Indian nationalism.

In Europe, steam power evolved gradually and uncertainly, with innovative peaks and long plateaus, from Thomas Savery's steam pump (1698) to James Watt and Matthew Boulton's double-acting rotative steam engine with a separate condenser (1765–1790). For most of the eighteenth century, steam power was provided by Thomas Newcomen's reciprocating steam pump (1712), which served primarily to remove water from coal mines. Watt and Boulton radically improved upon Newcomen's engine, eventually transforming it from a reciprocating pump into a source of rotative power for other technological innovations. The Watt engine literally and figuratively joined the jenny (1768), the water frame (1769), the boring mill (1775), the mule (1779), the seed drill (1782), cotton printing machinery (1783), the rolling mill for puddling iron (1784), the threshing machine (1786), and the improved lathe (1794) (Hartwell, 1967, pp. 69–70), as well as boats, ships, and trains after the turn of the new century, in launching a "coal- and steam-powered revolution in production, transportation, and extraction" (Adas, 1989, p. 136).

In India, the public history of steam power begins with three remarkable scenes, to each of which we will return.[1] First, just north of Calcutta in the Danish colony of Serampore (Srirampur), on March 27, 1820, the Baptist missionaries staged a demonstration of their new engine imported from Messrs. Thwaites and Rothwell of Bolton to assist with paper production for the mission press: "The 'machine of fire,' as they called it, brought crowds of natives to the mission, whose curiosity tried the

---

[1] Reports of earlier steam engines in use in India, prior to 1820, include the following. Discussing A. Couchlane's "interesting retrospect of the ship-building industry in India," *The Indian Review* for June 1916 mentions that "The first steam vessel used on the Indus – the *Snake* – was designed and built by a Parsee in 1800" ("Ship-Building," 1916, p. 429). W.H. Carey mentions a steamboat, the *John Shore*, launched from Kidderpore in 1807, adding, "But we have failed to find any account of her performances" (Carey, 1882, vol. 2, p. 18). Tann and Aitken record that the first installation of a Boulton and Watt engine was ordered in 1807 for use as a pump at the dockyard in Bombay, that the engine and a mechanic were sent out in 1809 (Tann and Aitken, 1992, p. 206), and that the firm of Fawcett and Littledale (of Liverpool) "supplied an engine to Calcutta in 1818" (p. 212). Blair Kling writes that "The first steam engine had been shipped out to Calcutta from Birmingham in 1817 or 1818. It was purchased by the government, fitted with buckets, and used to dredge the Hooghly River" (Kling, 1976, pp. 64–65), but he does not provide a source, and it is unclear if and when the engine arrived and went into operation. And in *An Account of Steam Vessels and of Proceedings Connected with Steam Navigation in British India* (1830), G.A. Prinsep reports that a private boat was built by William Trickett in 1819 in Lucknow for the Nawab of Oude (Awadh): "This little steamer, a toy of his late Majesty of Oude, was, like other playthings, soon laid aside, and suffered to go to ruin" (Prinsep, 1830, p. 3). According to Prinsep, the first engine to be "brought to Calcutta" arrived in 1817 or 1818, but it "remained some years neglected in a godown" until it was purchased to be fitted in a dredging boat, the *Pluto*, in 1822 (p. 2). Prinsep is much closer in time to the events he describes, but W.S. Lindsay in 1876 provides a contradictory account, claiming that the steamboat was built in Batavia in 1810–1811, named the *Van der Capellen*, and later brought to Calcutta, renamed the *Pluto*, and repurposed as a dredging boat in 1822 (Lindsay, 1876, p. 448).

patience of the engineman imported to work it; while many a European who had never seen machinery driven by steam came to study and to copy it" (G. Smith, 1885, p. 245).[2] The second high-profile stationary engine in Bengal went into operation two and a half years later, on November 1, 1822, in Calcutta at Chandpal Ghat to supply water from the Hugli river via aqueducts to the streets of the city's "White Town" in order to keep down the dust (Carey, 1882, vol. 1, p. 59). In the following summer, "At exactly nine minutes past four on Saturday afternoon (12th July) the first steam vessel, which ever floated on the waters of the East," the Diana Packet, "left the stocks at Kyd's Yard, Kidderpore" ("Launch," 1824, p. 195), with a more public launch following on the morning of August 9, when the Diana departed from Chandpal Ghat, "stemming the rapid freshes of the river with a velocity perfectly astonishing" ("The Diana," 1824, p. 280). The first steamer to complete the journey from England to India, the *Enterprize* (an "auxiliary" paddle ship equipped with sails), left Falmouth on August 16, 1825 and arrived in Calcutta via the Cape of Good Hope a disappointing 113 days later, of which (due to bad weather and exhausted fuel) only 62 were under steam (Hoskins, 1926, pp. 114–15).[3] Between 1822 and 1837, the number of steam engines working in Bengal rose from 2 to 67, and by 1845, according to *The Bengal Hurkaru*, there were 150 engines in use (mostly imported), roughly half of which (73) powered ships and boats, both sea and inland, with the largest groups of the remainder running sugar mills (28), raising water levels at the docks (8), removing water from coal mines (8), and driving the mint and other government departments (8), flour and rice mills (6), and paper factories (4) (Kling, 1976, p. 65). By the end of this period, in 1844, J.H. Stocqueler could report that "On approaching Calcutta, the smoking chimneys of steam-engines are now seen in every direction, on either side of the river, presenting the gratifying appearance of a seat of numerous extensive manufactories, vying with many British cities" (Stocqueler, 1845, p. 348).

---

[2] When Fathullah Khan, son of Shah Lotf Ali Khan, paid a state visit in October 1822, the Marquess of Hastings received him at the Governor-General's summer residence in Barrackpore, right across the river from Serampore: "During his Highness's stay there his Lordship gratified him with a sight of the steam engine, which seemed to excite his admiration, and to afford him much pleasure" ("The Persian Prince," 1823, p. 414).

[3] See as well Lester et al. (2021, pp. 169–70).

As Wordsworth predicted in 1800, in the Preface to the second edition of *Lyrical Ballads*, "If the time should ever come when what is now called Science, thus familiarized to men, shall be ready to put on, as it were, a form of flesh and blood, the Poet will lend his divine spirit to aid the transfiguration" (Wordsworth and Coleridge, 2013, p. 107). As we will see, the steam engine did put on "a form of flesh and blood" in ways that may seem surprising to readers of Romantic literature, and Wordsworth's prediction came true. Joanna Baillie wrote a profoundly ambivalent "Address to a Steam Vessel" (1823), while S.T. Coleridge, Byron, P.B. Shelley, and Thomas Love Peacock were all fascinated by steam power. Ute Berns has recently conducted a daring reading of Coleridge's "Rime of the Ancient Mariner" (1798) as a sustained engagement with "the principles and chemistry of heat, air and water considered to be at the core of the steam engine" (Berns, 2023, p. 37).[4] In *Don Juan*, Byron describes his need to rhyme as the "steam-boat which keeps verses moving" (Byron, 1986, p. 431), and Andrew Barbour has persuasively shown how Byron "develops a mechanical engineering poetics" (Barbour, 2021, p. 114). Having contemplated setting up a steam-driven grain mill in Tuscany and bankrolled a failed scheme in 1819–1820 to build a steamboat that would connect Leghorn, Marseille, and Genoa, in "A Letter to Maria Gisborne" (1820) Shelley correlates the "devilish enginery" of his writing with "The self-impelling steam-wheels of the mind" (Shelley, 2011, p. 445).[5] And in *Crotchet Castle* (1831) the Society for the Diffusion of Useful Knowledge, founded in 1826, was nicknamed "The Steam Intellect Society" by Peacock, who served as a star witness before the House of Commons' Select Committee on Steam Navigation to India in 1834 and succeeded James Mill as the East India Company's Chief Examiner of Indian Correspondence in 1836.[6] As early as 1814, Wordsworth himself put a (not unambiguous) paean to the "potent Enginery" of "An inventive Age" (Wordsworth, 2007, p. 261) in the mouth of the Wanderer in Book VIII of *The Excursion*. In *Yarrow Revisited* (1835), among the "Poems composed during a Tour in Scotland, and on the English Border, in the Autumn of 1831," he then

---

[4] See as well Scott (2018, pp. 242–43). Coleridge memorably referred to Thomas Clarkson as "the Moral Steam-Engine" (S.T. Coleridge, 1959, p. 179).

[5] See Allen (1988), Jones (2019), and Speitz (2019).

[6] See Headrick (1981, pp. 23–37).

included "Steamboats, Viaducts, and Railways," "The Pibroch's Note" (which lists "The smoking steam-boat" among "the conquests of civility" [Wordsworth, 2004, p. 498]), "On the Frith of Clyde (in a Steamboat)," and "Stanzas Suggested in a Steamboat off Saint Bees' Heads, on the Coast of Cumberland." In 1832 he added six lines celebrating steamships to his 1822 poem, "To Enterprize," and in 1844 he issued a NIMBY sonnet in *The Morning Post*, "On the Projected Kendal and Windermere Railway," in which the poet calls on the voice of nature to join his own in protesting the wrongs of Lake District tourism. Far less well known than Wordsworth, the "Corn-Law Rhymer" Ebenezer Elliott published the fascinating "Steam—a Poem" in *The New Monthly Magazine* for 1833, and his "On the Opening of the Sheffield and Rotherham Railway" followed in *Tait's* in 1838.[7] To round out this incomplete survey (there is an enormous amount of "railway poetry" starting in the 1830s), steam figures prominently in Thomas Carlyle's "Age of Machinery" in *Signs of the Times* (1829) and Thomas De Quincey's *The English Mail Coach* (1849).

While Carlyle was admonishing the readers of *The Edinburgh Review* that "Our grand business undoubtedly is, not to *see* what lies dimly at a distance, but to *do* what lies clearly at hand" (Carlyle, 1869, p. 313), anglophone writers in Calcutta were relentlessly looking forward and projecting utopian or dystopian futures powered by steam, and part of my argument will be that by seeing what lay dimly at a distance, they were in fact doing what lay clearly at hand—fashioning the politics of empire in concert and conflict with liberalism, Romanticism, and utilitarianism. And they were doing so during the "watershed" moment that science fiction scholar Darko Suvin places "around 1800, when space loses its monopoly upon the location of estrangement and the alternative horizons shift from space to time" (Suvin, 1979, p. 89).[8] As speculative fictions, in other words, these futures were fantastical realizations and projections of the very conditions of liberal imperialism, assimilation, hybridity, and

---

[7] "Steam—a Poem," published in *The New Monthly Magazine* for May 1833, is better known under its later title, "Steam, at Sheffield," as it appeared in Elliott's *Kerhona, The Vernal Walk, Win Hill, and Other Poems* (1835).

[8] Suvin defines science fiction as "*a literary genre whose necessary and sufficient conditions are the presence and interaction of estrangement and cognition, and whose main formal device is an imaginative framework alternative to the author's empirical environment*" (Suvin, 1979, pp. 7–8). On the shift from space to time in nineteenth-century futurist fiction, see Alkon (1987, pp. 3–88).

movement that structured the colonial present in the 1830s. David Lester Richardson's *The Bengal Annual* for 1835 included two such works, on which the final part of this Pivot will focus. "1980" by Henry Hurry Goodeve (1807–1884) introduces its characters "of all colors and ranks" as passengers on "the Himalaya steam mail" (fitted with "air-cooling machines") (Goodeve, 1835, p. 147), depicts Calcutta's Hugli river full of "steam dinghies innumerable ... skimming about in all directions" (p. 150), and features a dramatic escape from prison in "a steam balloon!" (p. 157). And in "The Junction of the Oceans. [*A Tale of the Year 2074.*]" by Henry Meredith Parker (1795–1863), "immense engines" (Parker, 1835, p. 12) cut a vast canal across Panama to let "gigantic" steamships (p. 32) navigate between the Atlantic and Pacific, inadvertently unleashing a flood that destroys the world.[9] A year before, Parker, Chairman of the New Bengal Steam Fund (constituted in June 1833, with Dwarkanath Tagore among the Directors), had published his "Thoughts on Flying" in Richardson's *The Calcutta Literary Gazette* for January 1834, a revery in which "Steam appears to be considered as the agent which holds out the most promising hope to the human race of being enabled to exchange the beaten ways of earth for the invisible paths of air" (Parker, 1834, p. 3), a hope that would then be graphically fictionalized in Émile Souvestre's *Le monde tel qu'il sera* [*The World as It Shall Be*] (1846; see Fig. 1.1).[10] And at the end of 1834 the *Gazette* printed "A Family Conversation," "*Scene*—CALCUTTA. *Date*—1935 A.D.," by an anonymous writer ("I"), in which the family's patriarch, machine-crazed Raj Kishen Roy, reads the news of the day, which includes a military engagement between the "First Consul of Spitzbergen and the Empress of the Esquimaux country" in which "The Lord High Admiral of the Esquimaux ship" is rendered

---

[9] Both Goodeve's "1980" and Parker's "The Junction of the Oceans" have recently been republished in Gibson (2019). For his collection *Bole Ponjis* (1851), Parker revised the tale and changed the subtitle to "A Tale of the Year 2098." Gibson's edition relies on this revised version (Parker, 1851, pp. 132–215). For a list of changes, see Gibson (2019, p. 36). For the sake of historical context, I quote throughout from the tale originally published in *The Bengal Annual* for 1835. Several sources, including Gibson, give the years of his birth and death as "1796?–1868"; he was born on June 4, 1795 ("The Humble Petition," 1814, p. 21) and died on September 17, 1863 ("High Court," 1864, p. 337), with his obituary appearing two days later in *The Homeward Mail from India, China, and the East* ("Henry Meredith Parker," 1863).

[10] On balloon flight, see Holmes (2008, pp. 125–62), and on ways in which new aerial technologies (especially balloons) shaped the Romantic imagination, see Gilroy (2022).

M. John Progrés.

Fig. 1.1 M. John Progrés "commodément assis sur une locomotive anglaise" [Mr. John Progress "comfortably seated on an English engine"], from Émile Souvestre, *Le monde tel qu'il sera* (1846, p. 9)

"helpless from the explosion of the boiler" ("A Family," 1834, p. 354). Set in the future, these pieces prophetically saw technology in general and steam in particular as the means by which the nineteenth, twentieth, and twenty-first centuries (most of our power plants today, whether fueled by coal, natural gas, nuclear fission, or solar thermal energy, generate electricity by means of steam turbines) would be transformed into an age of postcolonialism, neocolonialism, and global capitalism, a future world to which, the writers for the Indian periodicals realized, their own place and time were in the process of giving birth.

My argument and intervention, however, lie not in the prescience of these speculative fictions—Goodeve's techno-optimism in particular failed spectacularly as a political prediction about the future of Indian independence and nationalism, especially when compared with Kylas Chunder Dutt's "A Journal of 48 Hours of the Year 1945" (1835)—but in the ways they unevenly map the politics of liberal imperialism on to attitudes toward steam power and technology. It is no accident that these fantasies were clustered in the mid-1830s. The 1813 renewal of the East India Company's Charter had already ended the Company's monopoly, authorized missionary activity within Company territories, and funded English-language education in Western sciences for natives of India, all victories for the free-trade interest and the agenda of anglicization with which it was allied over the old "orientalist" order dating back to Warren Hastings, William Jones, and Henry Thomas Colebrooke.[11] The 1833 Charter then opened British India almost entirely to free trade, dissolving the Company's commercial mandate altogether and transforming it from a dual governing and trading company into a purely administrative ruling body (which retained its army and the right to wage war), and for the first time legalized colonization by Britons, who had hitherto been forbidden from settling without license from the Company and from owning land.[12] In and around 1835, in other words, under William Bentinck as the first "Governor-General of India" and Thomas Babington Macaulay as a member of the newly formed Governor-General's Council, and with British India in the grips of an economic crisis following the collapse of the agency houses between 1830 and 1834, writers imagining the present would have felt as if they were embarking upon an uncertain and unknown future inaugurated by the demise of an old order and the triumph of a new—a triumph, in short, of liberalism.

As Javed Majeed has shown, in the wake of James Mill's *The History of British India* (1817), "different strands of thought in British intellectual circles engaged with each other as they grappled with the host of political and philosophical problems which were implicit in the growth of British power in India" (Majeed, 2005, p. 93). C.A. Bayly (2012) and Partha Chatterjee (2012) have disagreed over the extent to which there was a continuous liberal tradition beyond British circles, among Indian intellectuals, with Chatterjee suggesting a rupture between the "historical formation" of anti-absolutism in the early nineteenth century and later colonial models of nationalism and modernity. Chatterjee locates this historical formation in the writings and activities of three groups: "nonofficial Europeans, Eurasians, and members of the new Indian elite"

---

[11] The funding for education never materialized, with only two years paid in arrears in 1824 ("Minutes," 1852–53, p. 247).

[12] On the colonization question, see Chaudhuri (2018).

(Chatterjee, 2012, p. 157). Although it is true that nonofficial Europeans spearheaded the liberal agenda in their opposition to Company power or "despotism," Company servants such as Goodeve, Parker, and others with roles to play in the story I am going to tell were pulled in different directions by the tensions and contradictions within liberal thought, especially as it dovetailed with utilitarianism in the 1830s.[13] The goal of this historicist literary study is not to fit literary history into the history of ideas or political theory but rather to show how divisions within the culture of liberal imperialism, informed by but by no means always faithful to the ideas of Jeremy Bentham, Macaulay, and the Mills, and representations of steam power were mutually constitutive for creative writers who were also servants of empire. This Pivot first focuses on a Romantic strain of thought and sentiment according to which steam technology represents an anti-utilitarian humanization of nature and then places within and against that frame the uneven and divergent ways in which writers in British India map a constellation of liberal values—support for free trade, the freedom of the press, colonization, and assimilation or "anglicization," including the replacement of Persian with English as the language of administration, advocacy for native education in English, and a program of Christianization—on to their hopes and fears concerning a future powered by steam.

References

Adas, Michael. (1989). *Machines and the Measure of Men*. Ithaca: Cornell University Press.

Alkon, Paul K. (1987). *The Origins of Futuristic Fiction*. Athens, GA: University of Georgia Press.

Allen, Barbara T. (1988). "Poetry and Machinery in Shelley's 'Letter to Maria Gisborne'." *Nineteenth-Century Studies*, 2, 53–61.

Barbour, Andrew. (2021). "The Rise of Thermodynamics: Mechanical Engineering and Byron's Poetic Machinery." *ELH*, 88(1), 107–31.

Bayly, C.A. (2012). *Recovering Liberties: Indian Thought in the Age of Liberalism and Empire*. Cambridge: Cambridge University Press.

Berns, Ute. (2023). "Anthropocene Speculations: Steam Technology in Coleridge's 'Rime of the Ancient Mariner' (1798)." *European Romantic Review*, 34(1), 19–46.

---

[13] My thinking on liberal imperialism is primarily informed by Mehta (1999), Pitts (2005), Sartori (2008), and Koditschek (2011).

Blackmantle, Bernard [pseud. Charles Molloy Westmacott]. (1826). *The English Spy: An Original Work, Characteristic, Satirical, and Humorous*. Vol. 2. London: Sherwood, Gilbert, and Piper.

Byron. (1986). *Lord Byron: The Complete Poetical Works, Vol 5: Don Juan*. Ed. by Jerome J. McGann. Oxford: Clarendon Press.

Carey, W.H. (1882). *The Good Old Days of Honorable John Company*. 2 vols. Simla: Argus Press.

Carlyle, Thomas. (1869). "Signs of the Times. [1829.]" *Critical and Miscellaneous Essays: Collected and Republished*. Vol. 2. London: Chapman and Hall, pp. 313–42.

Chatterjee, Partha. (2012). *The Black Hole of Empire: History of a Global Practice of Power*. Princeton: Princeton University Press.

Chaudhuri, Rosinka. (2018). "'On the Colonization of India' (1829): Public Meetings, Debates and Later Disputes." *The Indian Economic and Social History Review*, 55(4), 463–89.

Cohn, Bernard. (1996). *Colonialism and its Forms of Knowledge: The British in India*. Princeton: Princeton University Press.

Coleridge, Samuel Taylor. (1959). *The Collected Letters of Samuel Taylor Coleridge, Vol. 3: 1807–1814*. Edited by Earl Leslie Griggs. Oxford: Oxford University Press.

De Quincey, Thomas. (2003). "The English Mail-Coach, or the Glory of Motion." In Robert Morrison, ed., *The Works of Thomas De Quincey, Vol. 16: Articles from* Tait's Edinburgh Magazine, MacPhail's Edinburgh Ecclesiastical Journal, *the* Glasgow Athenaeum Album, *the* North British Review, *and* Blackwood's Edinburgh Magazine, *1847–1849*. London: Pickering & Chatto, pp. 408–28.

"The Diana Steam Packet." (1824). *Asiatic Journal*, 7(99), 280–81.

Dutt, Kylas Chunder. (1835). "A Journal of 48 Hours of the Year 1945." In D.L. Richardson, ed., *The Calcutta Literary Gazette or Journal of Belles Lettres, Science, and the Arts*, new series 3(75) (June 6), 355–58.

Elliott, Ebenezer. (1833). "Steam—A Poem." *The New Monthly Magazine*, 38(149), 30–34.

———. (1835). "Steam, at Sheffield." *Kerhona, The Vernal Walk, Win Hill, and Other Poems*. London: Benjamin Steill.

"A Family Conversation." (1834). In D.L. Richardson, ed., *The Calcutta Literary Gazette or Journal of Belles Lettres, Science, and the Arts*, new series 2(49) (December 6), 353–56.

Gibson, Mary Ellis, ed. (2019). *Science Fiction in Colonial India, 1835–1905: Five Tales of Speculation, Resistance and Rebellion*. London: Anthem Press.

Gilroy, John. (2022). *Romantics and the Era of Early Flight*. Cham, Switzerland: Palgrave Macmillan.

Goodeve, Henry Hurry. (1835). "1980." In David Lester Richardson, ed., *The Bengal Annual. A Literary Keepsake for MDCCCXXXV*. Calcutta: Samuel Smith, pp. 147–83.

Hartwell, R.M. (1967). *The Causes of the Industrial Revolution in England*. London: Methuen.

Headrick, Daniel R. (1981). *The Tools of Empire: Technology and European Imperialism in the Nineteenth Century*. New York: Oxford University Press.

"Henry Meredith Parker." [Obituary.] (1863). *The Homeward Mail from India, China, and the East*, September 19, 801–802.

"High Court of Judicature at Fort William in Bengal. Testamentary and Intestate Jurisdiction in the Good of Henry Meredith Parker deceased." [Will.] (1864). British Library India Office Records, Bengal Wills (February 19, 1864), L/AG/34/29/109/336–48.

Holmes, Richard. (2008). *The Age of Wonder: How the Romantic Generation Discovered the Beauty and Terror of Science*. New York: Pantheon Books.

Hoskins, Halford L. (1926). "The First Steam Voyage to India." *Geographical Review*, 16(1), 108–16.

"The Humble Petition of Henry Meredith Parker." (1814). British Library India Office Records J1/29/14–22 1814.

Jones, Steven E. (2019). "Shelley's 'Letter to Maria Gisborne' as Workshop Poetry." *The European Legacy*, 24(3–4), 380–95.

Kling, Blair B. (1976). *Partner in Empire: Dwarkanath Tagore and the Age of Enterprise in Eastern India*. Berkeley: University of California Press.

Koditschek, Theodore. (2011). *Liberalism, Imperialism, and the Historical Imagination: Nineteenth-Century Visions of a Greater Britain*. Cambridge: Cambridge University Press.

"Launch of a Steam Packet." (1824). *Asiatic Journal*, 7(98), 195.

Lester, Alan, Kate Boehme, and Peter Mitchell. (2021). *Ruling the World: Freedom, Civilisation, and Liberalism in the Nineteenth-Century British Empire*. Cambridge: Cambridge University Press.

Lindsay, W.S. (1876). *History of Merchant Shipping and Ancient Commerce*. Vol. 4. London: Sampson Low, Marston, Low, and Searle.

Majeed, Javed. (2005). "James Mill's *The History of British India*: The Question of Utilitarianism and Empire." In Bart Schultz and Georgios Varouxakis, eds., *Utilitarianism and Empire*. Lanham, MD: Lexington Books, pp. 93–105.

Mehta, Uday Singh. (1999). *Liberalism and Empire: A Study in Nineteenth-Century British Liberal Thought*. Chicago: University of Chicago Press.

"Minutes of Evidence Taken before Select Committee on the Government of Indian Territories" [June 30, 1853]. (1852–53). *The Sessional Papers Printed by Order of the House of Lords*, 19, 221–53.

Parker, Henry Meredith. (1834). "Thoughts on Flying." In David Lester Richardson, ed., *The Calcutta Literary Gazette or Journal of Belles Lettres, Science, and the Arts*, new series 1 (January 4), 3–4.

———. (1835). "The Junction of the Oceans. [*A Tale of the Year 2074.*]." In David Lester Richardson, ed., *The Bengal Annual. A Literary Keepsake for MDCCCXXXV*. Calcutta: Samuel Smith, pp. 1–55.

———. (1851). *Bole Ponjis. Containing The Tale of the Buccaneer; A Bottle of Red Ink; The Decline and Fall of Ghosts; and Other Ingredients*. Vol. 1. London: W. Thacker.

Peacock, Thomas Love. (1831). *Crotchet Castle*. London: T. Hookham.

"The Persian Prince Futteh Oolla Khan." (1823). *The Asiatic Journal*, 15(88), 412–14.

Pitts, Jennifer. (2005). *A Turn to Empire: The Rise of Imperial Liberalism in Britain and France*. Princeton: Princeton University Press.

Prinsep, G.A. (1830). *An Account of Steam Vessels and of Proceedings Connected with Steam Navigation in British India*. Calcutta: Government Gazette Press.

"Reflections of a Returned Exile." (1837). *Asiatic Journal*, 22(1), 65–75.

Sartori, Andrew. (2008). *Bengal in Global Concept History: Culturalism in the Age of Capital*. Chicago: University of Chicago Press.

Scott, Heidi C.M. (2018). *Fuel: An Ecocritical History*. London: Bloomsbury Academic.

Shelley, Percy Bysshe. (2011). *The Poems of Shelley, Volume 3: 1819–1820*. Edited by Jack Donovan, Cian Duffy, Kelvin Everest, and Michael Rossington. Abingdon: Routledge.

"Ship-Building in India." (1916). *The Indian Review*, 17(6), 429.

Smith, George. (1885). *The Life of William Carey, D.D. Shoemaker and Missionary. Professor of Sanskrit, Bengali, and Marathi in the College of Fort William, Calcutta*. London: John Murray.

Souvestre, Émile. (1846). *Le monde tel qu'il sera*. Paris: W. Coquebert.

Speitz, Michele. (2019). "Lyres, Levers, Boats, and Steam: Shelley's Dream of a Correspondent Machine." *Studies in Romanticism*, 58(2), 231–64.

"Steam Navigation." (1830). *Calcutta Magazine*, 5, 319–23.

Stocqueler, J.H. (1845). *The Hand-Book of India, a Guide to the Stranger and the Traveller, and a Companion to the Resident*. 2nd ed. London: Wm. H. Allen.

Suvin, Darko. (1979). *Metamorphoses of Science Fiction: On the Poetics and History of a Literary Genre*. New Haven: Yale University Press.

Tann, Jennifer, and John Aitken. (1992). "The Diffusion of the Stationary Steam Engine from Britain to India 1790–1830." *The Indian Economic and Social History Review*, 29(2), 199–214.

Wordsworth, William. (2004). *Sonnet Series and Itinerary Poems, 1820–45*. Edited by Geoffrey Jackson. Ithaca: Cornell University Press.

————. (2007). *The Excursion*. Edited by Sally Bushell, James A. Butler, and Michael C. Jaye. Ithaca: Cornell University Press.

Wordsworth, William, and Samuel Taylor Coleridge. (2013). *Lyrical Ballads 1798 and 1802*. Edited by Fiona Stafford. Oxford: Oxford University Press.

CHAPTER 2

A Soul Imparted to Brute Matter; or, the Secret Ministry of Steam

Abstract This chapter traces a surprising aestheticization of steam power from accounts of the second stationary engine in operation in Bengal through Erasmus Darwin's *The Botanic Garden* (1791) and eighteenth-century depictions of steam engines as living creatures to Wordsworth's ambivalent celebrations of technology in "Steamboats, Viaducts, and Railways" (1835) and *The Excursion* (1814). Along the way, I use Darwin's poem to assist with an accessible but detailed description of the inner workings of Thomas Savery's pump and Thomas Newcomen's engine (a note sketches James Watt and Matthew Boulton's improvements). Treating Wordsworth's poetry before ending with a reading of Coleridge's "Frost at Midnight," the chapter concludes that Romantic, anti-utilitarian accounts of engines as spontaneous and self-impelling, as vital, in-spirited, or even human, involve a considerable amount of mystification, resisting but ultimately failing to confront the utilitarian "productivist" (Rabinbach) calculus according to which tools that elide living workers and transform sited, embedded labor into abstract, mobile, and placeless production promote the greater good, perpetuate improvement, and advance civilization.

Keywords Steam power in Britain · William Wordsworth · S.T. Coleridge · Erasmus Darwin · Romanticism · Utilitarianism

© The Author(s), under exclusive license to Springer Nature Switzerland AG 2024
D. E. White, *Romanticism, Liberal Imperialism, and Technology in Early British India*, https://doi.org/10.1007/978-3-031-60705-9_2

On November 4, 1822, *The Bengal Hurkaru* newspaper announced that "On Friday evening, about sun-set, the beautiful steam engine erected at Chaundpaul Ghaut for watering the streets of Calcutta, was put in motion for the first time" ("Steam Engine," 1823, p. 524).[1] Surprisingly aestheticized, the description that followed of this second stationary steam engine in Bengal as neither dark nor Satanic suggested that its "Motions and Means" were not necessarily "at war / With old poetic feeling" (Wordsworth, 2004, p. 604), or at least, in this instance, with neoclassical taste: "The house which contains the engine, boiler, and pumps, is a neat regular octagon, in the Doric style, of 50 feet diameter inside. The exterior has an excellent effect, and the chimney (a chaste Doric fluted column) upwards of 70 feet high, rising from the centre, gives it more the air of an antient mausoleum than the receptacle of a steam engine … The angles are agreeably relieved by double fluted pilasters resting on a basement about four high, and supporting a chaste cornice appropriate to the order of the building" ("Steam Engine," 1823, p. 524). Also depicted in Robert Burford's Panorama of Calcutta, which was exhibited in Leicester Square in 1830, the engine in both the newspaper account and William Wood's 1833 panoramic plate (Fig. 2.1) is easily assimilated within a harmony of landscape, architecture, and technology that naturalizes imperial rule.[2] Power, here, be it mechanical, commercial, or political, is eminently chaste, proportionate, and tasteful.[3]

But the aestheticization of steam power went beyond neoclassicism. A distinctly Romantic, and ultimately anti-utilitarian, strain, in which the inorganic machine puts on a form of flesh and blood, can be heard as early as Part I of *The Botanic Garden* (1791), where Erasmus Darwin traces the progress of technology from Savery's steam pump through to Newcomen's and Watt's engines and their applications (water engines, corn mills, mints), ending with a near-future fantasy in which "Soon

---

[1] The article in *The Bengal Hurkaru* was reprinted in *The Asiatic Journal* for May 1823, from which I quote.

[2] On the relationship between the panoramic perspective and "overlordship," see Charlesworth (2001).

[3] Intriguingly, the shape of the octagonal housing with the chimney rising above strongly resembles that of Holwell's Monument to the Black Hole of Calcutta, an octagon with an obelisk, erected in 1760 on the northwest corner of Tank Square, not far to the north of Chandpal Ghat, and demolished in 1821, by which time the monument had fallen into disrepair. (The present marble monument, in the graveyard of St. John's Church, is a replica, originally erected by Lord Curzon in Tank Square in 1902 and then relocated to St. John's Church in 1940.) Perhaps the shape of the steam engine's housing, which would have been designed precisely as the Black Hole monument was being dismantled, was meant to signal imperial continuity under a new regime of power, technological rather than military.

Fig. 2.1 Detail from "Chandpaul Ghaut. Steam Engine. Supreme Court," by William Wood, from *A Series of Twenty-Eight Panoramic Views of Calcutta* (1833), Plate 1; from the British Library Archive shelfmark 1781.c.22

shall thy arm, UNCONQUER'D STEAM! afar / ... on wide-waving wings expanded bear / The flying-chariot through the fields of air" (Darwin, 1791, p. 29). His poetic depiction is as remarkable for its compression of technical detail as it is for its personifications. Darwin starts by praising Savery, inventor of a pistonless pump for draining coal mines, and then describes Newcomen's engine:

> Nymphs! You erewhile on simmering cauldrons play'd,
> And call'd delighted Savery to your aid;
> Bade round the youth explosive Steam aspire
> In gathering clouds, and wing'd the wave with fire;
> Bade with cold streams the quick expansion stop.
> And sunk the immense of vapour to a drop.—
> Press'd by the ponderous air the Piston falls

Resistless, sliding through it's [*sic*] iron walls;
Quick moves the balanced beam, of giant-birth,
Wields his large limbs, and nodding shakes the earth. (pp. 26–27)

Because, like Darwin and unlike most today, Romantic-era writers and readers tended to be keenly interested in and well informed about the inner workings of these machines—"subjects of the highest importance," Shelley called them (Shelley, 1964, p. 132)—a quick sketch, which could supplement diagrams and videos online, may help, and Darwin's lines, in fact, can assist with the illustration.

Although Savery's pump (see Fig. 2.2) did not work for its intended purpose, its reliance on the capacity of water to expand from liquid into steam, to displace air, and then to condense back into liquid, producing a partial vacuum, paved the way for Newcomen's and Watt's adaptations. It operated by means of a chamber (E) attached on the one side, at its top, to a boiler (B, in which the "wave" of liquid water was "wing'd … with fire" by coal, A, burning below), and on the other side, at its bottom, to a joint (I-K) connecting a down-pipe (G), which descended into the water to be drained (H), and an up-pipe (L), which split in two above the chamber, first with an outlet (M) pouring water over the chamber and then an exhaust pipe (O) releasing the pumped water. There were four valves (also called taps, gauges, or plugs), two of which were opened and closed by hand—the boiler valve (C) and the up-pipe outlet valve (M)—and two by pressure—the down-pipe valve (I, at the bottom of the joint) and the main up-pipe valve (K, at the top of the joint). When the boiler valve was opened, "explosive" steam would fill the chamber, displacing the air inside ("the quick expansion"), and then the valve would be closed. The up-pipe outlet valve would then be opened (in this description, assume that the pump has been operating and the up-pipe already has water in it), pouring cool water over the chamber ("Bade with cold streams the quick expansion stop"): the steam in the chamber would condense back into water, shrinking in volume by a ratio of approximately 1,700:1 ("And sunk the immense of vapour to a drop"), thus forming a partial vacuum, and the up-pipe outlet valve would be closed. The resulting negative pressure differential (suction) would then open the down-pipe valve and draw the water up the down-pipe and into the chamber. The boiler valve would next be opened again: steam would fill the chamber above the water inside it, pushing the water out of the bottom of the chamber, through the up-pipe (opening the valve on its

Fig. 2.2 Savery's
Steam Pump, from John
Scott Russell, *A Treatise
on the Steam-Engine*
(1846, p. 31)

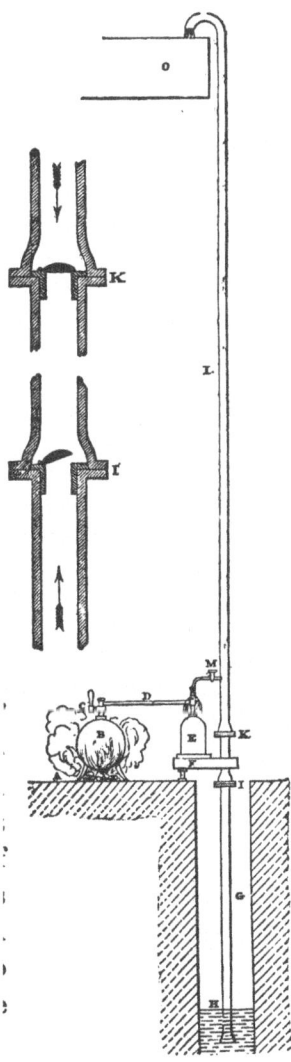

way), and out. (The steam had to be relatively high-pressure to perform this work, an "alarming" proposition given the state of late-seventeenth-century metallurgy [Cardwell, 1963, p. 16].) Once the chamber was empty of water and again full of steam, the up-pipe outlet valve would again be opened, pouring cool water over the chamber and condensing the steam inside it, and the process would then be repeated.

In 1712, Newcomen essentially replaced the chamber with a cylinder and added a piston mechanism inside it (see Fig. 2.3). The top of the piston (C) was exposed to the atmosphere ("the ponderous air"), and the piston was affixed by a rod and chain (D) to one side of a beam (F), which was connected on the opposite side (K) to the pump (M) from the mine (f) and a separate pump (k) to cycle water back to a cistern (Z). The beam was heavier on the pump side (K) than on the piston side (F). On the pump side, the reciprocal movement of the beam operated the pumps, and later, on the piston side, it also opened and closed the machine's own valves by means of pegs on a "plug tree" (d, a valve-gear mechanism added by Thomas Beighton in 1718), which ascended and descended alongside the piston rod, replacing the "plug man" (or boy) and transforming the pump into a true engine, capable of running by its own agency so long as its fuel lasted.[4] In Newcomen's engine, with the pump side of the beam up and the piston at the bottom of the cylinder, gravity would begin to pull the (heavier) pump side of the beam down: as this happened, the plug tree (details not shown in the diagram) on its way up would open the boiler valve (T), and, as gravity continued to pull the pump side of the beam down and raise the piston, steam would fill the cylinder under the piston, expelling the air through a "snifter valve" (f). The plug tree would rise along with the piston, and when the cylinder was full of steam the tree would then close the boiler valve, cutting off the steam, and open a water valve (S), spraying a stream of cold water (R) from the cistern directly into the cylinder under the piston. The cold water would condense the steam (which would drain through an education pipe, e, into a well, h) and thereby create a partial

---

[4] On the legendary origin of this innovation by "a boy who wanted to save his own labour" (Humphry Potter), see Adam Smith: "In the first fire-engines, a boy was constantly employed to open and shut alternately the communication between the boiler and the cylinder, according as the piston either ascended or descended. One of those boys, who loved to play with his companions, observed that, by tying a string from the handle of the valve, which opened this communication, to another part of the machine, the valve would open and shut without his assistance, and leave him at liberty to divert himself with his play-fellows" (A. Smith, 1993, p. 17).

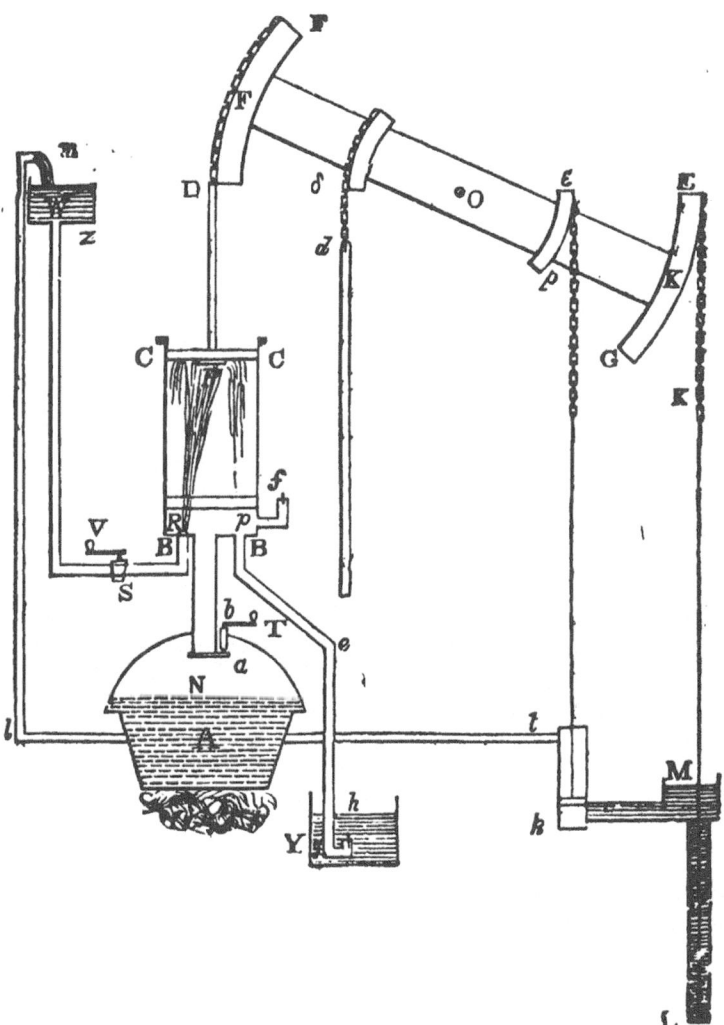

Fig. 2.3 Newcomen's Engine, from Russell, *A Treatise on the Steam-Engine* (1846, p. 50); the diagram shows the plug tree (d) but not the interface between its pegs and the valves

vacuum below the piston. With only a partial vacuum now beneath the piston, the atmospheric pressure above it (about 14 psi at sea level) would then push it down (the sole power stroke in a "single-acting" engine; "Press'd by the ponderous air the Piston falls"), thus lowering the piston side of the beam and raising the pumps on the opposite side ("Quick moves the balanced beam"). At its height, as gravity acting on the greater mass of the pump side of the beam then caused it to descend, the plug tree (now near its bottom and on its way up) would open the boiler valve while the beam would let the pump drop on the one side and would raise the piston on the other: the chamber below the piston would fill with steam once again as the piston rose, and the cycle would repeat. The engine thus worked by means of gravity for the piston's upstroke and atmospheric pressure above a partial vacuum on the downstroke. Because the steam was not doing work, as it was in Savery's pump (in which it pushed the water out of the chamber), it could be low pressure. Newcomen's was thus a single-acting, reciprocating "atmospheric engine."[5]

---

[5] Watt and Boulton's innovations turned Newcomen's reciprocating engine, which was suited primarily to pumping water out of coal mines, into a more efficient, universal engine capable of powering other machines. Because Newcomen's engine condensed steam by means of a jet of water sprayed directly into the cylinder, during each cycle steam was inefficiently reintroduced into a just-cooled cylinder, in which it would continue to condense until the cylinder heated up again. To overcome this waste of energy, in 1765 Watt developed a separate jet-condenser connected to the cylinder by a pipe and valve. The separate condenser (patented in 1769) was kept cool in a basin of water, and the pressure inside the condenser was kept relatively low by means of an air pump (operated by the beam). When the valve connecting the cylinder to the condenser opened, the condenser would draw the steam from the cylinder, leaving a partial vacuum under the piston while allowing the cylinder itself to remain warm throughout the cycle. Inside the condenser, a jet of water would condense the steam, and the air pump would then remove the resulting water from the condenser and maintain the low pressure for the next stroke. The transformation of reciprocal into rotatory motion then involved five innovations, each of which essentially generated the need for the next. First, the obvious method of using the reciprocal motion of the beam to turn a wheel was by means of a crank, but because the crank mechanism had already been patented (by James Pickard), Watt developed the sun-and-planet gear mechanism (1781) to drive a flywheel. Second, the uneven timing of the single-acting engine (in which the downstroke was fast and the upstroke slow), while suitable for a pump, was inadequate for a wheel, which needed to turn smoothly. By sealing the cylinder (so that even relatively low pressure steam could do work against a partial vacuum), admitting steam into the cylinder both above and below the piston, and connecting both the top and bottom of the cylinder to the condenser, Watt introduced a second and even power stroke, creating a double-acting engine (1782). But because the link between the piston and the beam in a single-acting engine was a chain, which could pull the beam down but not push it up, a different connection was required. A simple

Darwin's account is remarkable for more than its precision, though. In it, the work performed by the "arm" of steam, specifically by the physical capacity of water to expand from liquid into vapour and to contract back again into liquid, is natural and embodied, and the lines attribute biological agency to the personified beam "of giant-birth," which autonomously "Wields his large limbs, and nodding shakes the earth." Paul A. Cantor writes, "it has become a critical commonplace that the Romantics rejected technology as an unnatural transformation of the human condition" (Cantor, 1993, p. 110), but in John Tresch's compelling phrase, the steam engine could be a "romantic machine," merging "Concepts of mechanism and organicism" (Tresch, 2012, p. 5). *The Botanic Garden* in fact captures a longstanding, Romantic feeling about steam engines dating back to the mid-eighteenth century. Clearly inspired by Newcomen's engine, Bernard Forêt Bélidor's *Architecture hydraulique* (1737–53) included the following striking passage, which would resurface frequently up to a century later:

> Il faut avouer que voilà la plus merveilleuse de toutes les Machines, & qu'il n'y en a point dont le Mécanisme ait plus de rapport avec celui des animaux. La chaleur est le principe de son mouvement; il se fait dans ses différens tuyaux une circulation, comme celle du sang dans les veines, ayant des valvules qui s'ouvrent & se ferment à propos; elle se nourrit, s'évacue d'elle-même dans des tems réglés, & tire de son travail tout ce qu'il lui faut pour subsister. (Bélidor, 1739, pp. 324–25)

The title page of Robert Stuart's *A Descriptive History of the Steam Engine* (1824) takes the passage as its epigraph, translated thus:

> It must be acknowledged that this is the most wonderful of all machines, and that nothing of the work of man approaches so near to animal life. Heat is the principle of its movement; there is in its tubes a circulation like that of the blood in the veins of animals, having valves which open and shut in proper periods; it feeds itself, evacuates such portions of its food as

---

rigid rod would not do because, while the piston needed to move directly up and down inside the cylinder, the end of the beam moved in an arc. These challenges led to Watt's parallel motion mechanism in 1784. Finally, to maintain the even speed of the piston, the centrifugal governor connected to a throttle valve (1788) was added, and the pressure gauge (1790) to monitor the pressure of the steam in the boiler followed.

are useless, and draws from its own labours all that is necessary to its own subsistence.

Opening with "STEAM! all powerful steam!" (Alderson, 1834, p. 1), M.A. Alderson's prize-winning essay for the London Mechanics' Institution in 1833, *An Essay on the Nature and Application of Steam*, incorporates and builds on the passage (without attribution), bringing it up to date with subsequent improvements by Watt and Boulton (the centrifugal governor and the pressure gauge):

> It is in the property which the steam-engine possesses of regulating itself, and providing for all its wants, that the great beauty of the invention consists. It has been said that nothing made by the hand of man approaches so near to animal life. Heat is the principle of its movement; there is in its tubes circulation, like that of the blood in the veins of animals, having valves which open and shut in proper periods; it feeds itself, evacuates such portions of its food as are useless, and draws from its own labours all that is necessary to its own subsistence. To this may be added, that they now regulate so as not to exceed the assigned speed, and thus do animals in a state of nature. That the safety-valves, like the pores of perspiration, open to permit the escape of superfluous heat in the form of steam. The steam-guage [*sic*], as a pulse to the boiler, indicates the heat and pressure of the steam within, and the motion of the piston represents the action and the power of which it is capable. The motion of the fluids in the boiler represents the expanding and collapsing of the heart; the fluid that goes to it by one channel is drawn off by another, in part to be returned when condensed by the cold, similar to the operation of veins and arteries. (pp. 43–44)

In Peacock's *Crotchet Castle* (1831), Mr. Chainmail, "a good-looking young gentleman ... with very antiquated tastes" who "holds that the best state of society was that of the twelfth century" (Peacock, 1831, p. 85), is thus actually out of step with this Romantic naturalization of technology in an "age of wonder" (Holmes, 2008) when he laments that "everything about us is as artificial and as complicated as our steam-machinery" (Peacock, 1831, p. 174). And equally so was De Quincey, who disparaged the "blind insensate agencies" of engines juxtaposed to "the dilated nostril, spasmodic muscles, and thunder-beating hoofs" of "the noblest amongst brutes": "on the new system of travelling, iron tubes and boilers have disconnected man's heart from the ministers of his locomotion"

(De Quincey, 2003, p. 417). On the contrary, if not quite a companionable form, the engine over and over appears as something close—a living, breathing creature, such as the "snorting little animal, which I felt rather inclined to pat," in Frances Kemble's memorable "ecstasies" over George Stephenson's Liverpool and Manchester Railway, which she rode in August 1830, shortly before it opened. In an early instance of the soon-to-be-ubiquitous "iron horse" conceit, Kemble provides an extended metaphor in which the feet of the "steam-horse" are its wheels, the legs its pistons, and the reins its handle, while the water it drinks is water and the oats, of course, are coal (Kemble, 1878, pp. 158–61).[6]

Distinct from other products of "man's art" in its vitality, the engine is also distinct from other forms of animal life "in a state of nature" by virtue of being our own "lawful offspring," to reintroduce the language of Wordsworth's sonnet. But Wordsworth's reconciliation with "poetic feeling" (if not with the "old" variety) stems from more than either harmony and proportion or the analogy between the steam engine and the animal body. In "Steamboats, Viaducts, and Railways," the initial rejection of technology as "at war / With old poetic feeling" inaugurates a train of thoughts that leads the poet to judge and celebrate "what in soul" technology is:

> MOTIONS and Means, on land and sea at war
> With old poetic feeling, not for this,
> Shall ye, by Poets even, be judged amiss!
> Nor shall your presence, howsoe'er it mar
> The loveliness of Nature, prove a bar
> To the Mind's gaining that prophetic sense
> Of future change, that point of vision whence
> May be discovered what in soul ye are.

[6] A letter to the editor of *The Times* of September 1822 (reprinted in *The India Gazette* on March 13, 1823), signed MEPHITICUS, asks, "Why should not steam horses ... be constructed?" and imagines reading "in time to come" of the Epsom Races: "Three horses started: Mr. Stokehole's *Explosion*, Mr. Ash's *Skyrocket*, and Mr. Coke's *Tinderbox*. The race was won with some difficulty by *Explosion*. *Skyrocket* blew up about 100 yards from the winning post: the remains of his rider have yet to be found" ("Steam; and the Strides," 1822, p. 3B). In "Thoughts on Flying," Parker pens a dialogue between a gentleman and his servant, beginning, "John, bring my flying horse to the door at two." As Lady Arabella Fitz Plantagenet prepares "for a morning's airing," her sister Jane then asks her, "Bella, love—did you ever walk?," to which she replies, "No dearest, but Ma did once" (Parker, 1834, p. 3).

In spite of all that beauty may disown
In your harsh features, Nature doth embrace
Her lawful offspring in Man's art; and Time,
Pleased with your triumphs o'er his brother Space,
Accepts from your bold hands the proffered crown
Of hope, and smiles on you with cheer sublime. (Wordsworth, 2004,
p. 604)

The poem was published in 1835, the year of the two futurist fictions in *The Bengal Annual* to which we will later turn, and of the Great Moon Hoax in New York and Edgar Allan Poe's "The Unparalleled Adventure of One Hans Pfaall," and a year after Félix Bodin's *Le roman de l'avenir* [*The Novel of the Future*] (1834), in which Bodin coined the phrase, "littérature futuriste" (Bodin, 1834, p. 26), and on which we will focus at the beginning of Chapter 4. The speculative sonnet turns on the realization that it is only from an imagined future "point of vision" that we can retrospectively discover "what in soul ye are" in the present.[7] As much as steamboats, viaducts, and railways mar the loveliness of present-day nature, the mind's "prophetic sense" can see from that future point that "in soul" technology is, as Cantor writes, "a new source of human power, which might in true Baconian fashion be applied to 'the relief of man's estate.' The Romantics hoped to humanize modern science and technology, to enlist them in the service of their utopian vision of the future of humanity" (Cantor, 1993, p. 114). It is thus a crown specifically of "hope" for a better future that Time, pleased with the triumphs of technology over Space, accepts from the very motions and means whose marring "presence" nonetheless prophecies "future change." In the distance from neoclassical proportion to body to "what in soul" technology actually was lay its humanity. And in its humanity, nature, including human nature, recognizes its own lawful offspring.

Similarly, in *The Excursion* Wordsworth's Wanderer acknowledges "the darker side / Of this great change," the "outrage done to Nature" (Wordsworth, 2007, p. 263), yet still exults "to see / An Intellectual mastery exercised / O'er the blind Elements; a purpose given, / A perseverance fed; almost a soul / Imparted—to brute Matter" (p. 264). In seeing technology as not just a brute material embodiment of intellectual

---

[7] On the moon-voyage genre, see Bennett (1983). On Bodin, see Alkon (1987, pp. 245–89).

mastery over nature—a utilitarian tool—but as itself animated by soul, spirit, or vitality, the Wanderer fully humanizes natural power as technology. As Ted Underwood writes, "The humanization of nature is one of the strongest impulses of Wordsworthian lyricism" (Underwood, 2005, p. 117), and "Industrial machines are not merely (as writers like Spence had argued) useful ways of abridging labor and 'working up' the material provided by agriculture—they are for Wordsworth an expression of nature's primary productive power" (121).[8] Or as Heidi C.M. Scott puts it, "By blending man's mechanical genius with the vast organic powers that animated the earth, industrial mechanisms became living, evolving beings that elevated civilization to, and above, nature's level" (Scott, 2018, p. 127). Sara Coleridge perhaps best captures this ethos, writing in 1836 that "Life is the steam of the corporeal engine; the soul is the engineer who makes use of the steam-quickened engine" (S. Coleridge, 1873, p. 164). The soul of the steam-quickened machine is a human being, the engineer, its user and creator.

In the period, then, there is an enthusiastic perspective on technology at odds with the instrumentalism of utilitarianism. Eric Schatzberg has traced "two sharply divergent traditions about the nature of technology," the "cultural approach," according to which "technology is imbued with human values and strivings in all their contradictory complexity," and the "instrumental approach," which insists "that technology is a mere instrument that serves ends defined by others" and "portrays technology as narrow technical rationality, uncreative and devoid of values" (Schatzberg, 2018, p. 3). Today "the cultural understanding of technology is definitely a minority view" (p. 4), for the long history from industrial dehumanization to climate change and now A.I. has cleaved humanity from technology, producing anxieties that machines are not just purely instrumental but work through alien and unknowable means and toward ends indifferent or even contrary to our collective interests. In 1823, however, Baillie could address a poem *to* a steam vessel in which she asked, "What is this power which thus within thee lurks, / And, all unseen, like a mask'd

---

[8] This is the crux of the debate between Adam Smith and David Ricardo initiated by Smith's claim that in manufactures "nature does nothing; man does all" (A. Smith, 1993, p. 217; *cf*. p. 397). Ricardo asks, "Does nature nothing for man in manufactures? Are the powers of wind and water, which move our machinery, and assist navigation, nothing? The pressure of the atmosphere and the elasticity of steam, which enable us to work the most stupendous engines—are they not the gifts of nature?" (Ricardo, 1951, p. 76n).

giant works?" and then answered on its behalf, "Ev'n that which gentle dames, at morning's tea, / From silver urn ascending, daily see / With tressy wreathings playing in the air, / Like the loos'd ringlets of a lady's hair" (Baillie, 1823, p. 260). Although unseen, the power is relational, intimately associated with human, specifically domestic, values, and feminized through comparison to a seemingly natural but thoroughly cultural part of a woman's body. Steam here exceeds "a simple human-nature dyad of oppositional actors," and in these distinctly Romantic images of technology and its agency we see what Michele Speitz describes in an important forthcoming study as "a less reductive understanding of the material world, where earthly scenes are not populated simply by dyads of protagonists and antagonists, but by a more complicated set of actors bearing out intersecting and overlapping agencies, contingencies, and aggregates" (Speitz, 2024 [forthcoming], n.p.).[9]

But Romantic, anti-utilitarian accounts of engines as spontaneous and self-impelling, as vital, in-spirited, or even human, involve a considerable amount of mystification, and Underwood reminds us that "Nature is not middle class, and middle-class production is not spontaneous" (Underwood, 2005, p. 87). First, the fascination with steam—"the expansion and condensation of fluids with a power equal to that of the lightning itself" (Somerville, 1834, p. 229)—tends to ignore the source of that power in stock fuel, in the energy (ultimately of the sun) captured in and released by wood or coal, which powers the series of actions and reactions ending with either tea or the rotatory energy of the flywheel that performs work. Next, such accounts do not just replace animate power (of humans and horses) but also elide the actual beings that continue to labor, transforming their ongoing veiled activity into abstract and interchangeable units of labor power. Indeed, Andreas Malm sees the rise of steam power not as the necessary consequence of improved mechanical efficiency and market forces but as an opportunity for capitalists to bring about precisely this transformation in modes of production and social relations (Malm, 2016). Simultaneously, these depictions remove the sitedness of work, as engines rely on impersonal properties of heat and water, not particular, natural forces: a water mill can only be built on a suitable section of a stream, where it is driven by the flow force of a body of water (the "prime mover") embedded in local community, culture, and

[9] I am grateful to Michele Speitz for letting me read and cite her forthcoming book in manuscript.

history; a steam mill, in the phrase of Sadi Carnot's *Reflections on the Motive Power of Heat* (1824), is a "universal motor" (qtd. in Rabinbach, 1990, p. 45) that can be erected anywhere (including centers of population) and appears to be spontaneously driven by powers seemingly "not circumscribed by land or inherited wealth" (Simpson, 2009, p. 75). And finally, such accounts celebrate the utopian implications of the first law of thermodynamics, which suggested that we inhabit a world in which "economic production and natural force" could be "names for a single power" (Underwood, 2005, p. 182), while neglecting the second, which, although yet to be articulated, was of course front and center in the daily challenges, and therefore minds, of engineers. Although, as David Philip Miller has shown in *James Watt, Chemist* (2009), Romantic-era thought about heat was grounded in chemistry rather than physics (Berns, 2023, p. 28), well before the articulations of the first and second laws of thermodynamics, in the 1840s and 1850s, respectively, early-nineteenth-century engineers were profoundly aware that energy, while never destroyed, is always wasted when it is converted from one form to another within any given system—"the rude truth that a great part of the power of any engine is expended in friction" (Barbour, 2021, p. 96).

The humanization of engines, accordingly, resists but ultimately fails to confront the utilitarian calculus according to which tools that elide living workers and transform sited, embedded labor into abstract, mobile, and placeless production promote the greater good, perpetuate improvement, and advance civilization. As much as Romantic accounts of steam power reject the mechanization of life by finding poetry in the technological synthesis of human and natural power, they nonetheless join utilitarian celebrations in contributing to what Anson Rabinbach calls productivism, "the belief that human society and nature are linked by the primacy and identity of all productive activity, whether of laborers, of machines, or of natural forces" (Rabinbach, 1990, p. 3). Exaltations of the vitality and apparently autonomous agency of the steam engine therefore manifest middle-class anxieties about that class' own new power of potentially endless, mobile, and undifferentiated production (and capital) liberated from agriculture, ownership, and the rooted particularity of local place.

At the dawn of the industrial nineteenth century, then, steam performed a secret, middle-class ministry, "unhelped by any wind." In the period, in fact, that phrase would far more commonly describe a steamboat by the broad light of day than frost at midnight. For Coleridge himself in "Youth and Age" (1828), those "trim skiffs, unknown of yore,

/ ... ask no aid of Sail or Oar" and "fear no spite of Wind or Tide!" (S.T. Coleridge, 2001b, p. 1012). For Baillie in 1823, the unassisted steamer was a "Rover at will on river, lake, and sea" (Baillic, 1823, p. 262) that asks "No leave ... of either wind or tide" (p. 260), while for Wordsworth in 1832, steamboats no longer dread "breathless calms" but "In never-slackening voyage go / ... slighting sails and scorning oars" (Wordsworth, 2004, p. 400). Frost and steam, of course, are different phases of the same thing: if frost is the phase of water produced by the loss of kinetic energy, which, through the process of deposition, transforms vapor into solid crystal, steam is the vapor phase produced by the gain of kinetic energy through boiling, an acceleration of the process of evaporation. The out-in-out form of Coleridge's "Frost at Midnight" mirrors the transformation of water as, in the poem's synaesthetic conclusion, it gains and loses energy across its three phases, melting from snow and ice into water "in the sun thaw" as "the eave-drops fall," evaporating from its liquid state into vapor as "the nigh thatch / Smokes," depositing from vapor into the crystalline form of frost and freezing from liquid back into solid, the "silent icicles, / Quietly shining to the quiet Moon" (S.T. Coleridge, 2001a, p. 456). In the movements of the film that "fluttered on the grate," of the "gentle breathings" of the infant's lungs, and ultimately of the poet's mind as it moves through memory from the present to the past and then imagines the future of the babe, who shall "learn far other lore, / And in far other scenes!" (pp. 454–55), the poem suggests that the invisible yet transformative transfer of kinetic energy, which is neither created nor destroyed, is the eternal language of an "intelligible" world, the "eternal language, which thy God / Utters, who from eternity doth teach / Himself in all, and all things in himself" (p. 456). In this poem's version of "the one life within us and abroad / Which meets all motion and becomes its soul" (p. 233) in "The Eolian Harp," no matter how much frost slows down the world, whether in the extreme silentness of winter at midnight or at the molecular level, energy is conserved, and nature remains radically in motion, both in the present actuality of the poet's mind, the film fluttering on the grate, the infant's respiration, and the passage of water across its phases, as well as in the eternal potential of future transfers. If "Frost at Midnight" is inspired by the low-energy state of water to imagine, through memory of the poet's own past, a future in motion, in which the babe will "wander like a breeze" (p. 456), other speculative imaginings—utopian and dystopian fantasies produced by servants of what Parker would insist on calling "The Empire of the

Middle Classes"—were inspired by the high-energy state of water, secretly condensing and working within cylinders.

REFERENCES

Alderson, M.A. (1834). *An Essay on the Nature and Application of Steam, with an Historical Notice of the Rise and Progressive Improvement of the Steam-Engine*. London: Sherwood, Gilbert, and Piper.

Alkon, Paul K. (1987). *The Origins of Futuristic Fiction*. Athens, GA: University of Georgia Press.

Baillie, Joanna, ed. (1823). *A Collection of Poems, Chiefly Manuscript, and from Living Authors*. London: Longman, Hurst, Rees, Orme, and Brown.

Barbour, Andrew. (2021). *Mechanical Powers: Engineering and Romantic Poetics in the Early Anthropocene*. [Doctoral Dissertation, University of California, Berkeley.] UC Berkeley Electronic Theses and Dissertations. https://eschol arship.org/uc/item/1rq5p20f.

Bélidor, Bernard Forêt. (1739). *Architecture hydraulique, ou l'art de conduire, d'élever et de ménager les eaux pour les différens besoins de la vie*. Vol. 2. Paris: Charles-Antoine Jombert.

Bennett, Maurice, J. (1983). "Edgar Allan Poe and the Literary Tradition of Lunar Speculation." *Science Fiction Studies*, 10(2), 137–47.

Berns, Ute. (2023). "Anthropocene Speculations: Steam Technology in Coleridge's 'Rime of the Ancient Mariner' (1798)." *European Romantic Review*, 34(1), 19–46.

Bodin, Félix. (1834). *Le roman de l'avenir*. Paris: Lecointe et Pougin.

Cantor, Paul A. (1993). "Romanticism and Technology: Satanic Verses and Satanic Mills." In Arther M. Melzer et al., eds., *Technology in the Western Political Tradition*, Ithaca: Cornell University Press, pp. 109–28.

Cardwell, D.S.L. (1963). *Steam Power in the Eighteenth Century: A Case Study in the Application of Science*. London: Sheed & Ward.

Charlesworth, Michael. (2001). "Subverting the Command of Place: Panorama and the Romantics." In Peter J. Kitson, ed., *Placing and Displacing Romanticism*, Aldershot: Ashgate, pp. 129–45.

Coleridge, Samuel Taylor. (2001a). *The Collected Works of Samuel Taylor Coleridge: Poetical Works I: Poems (Reading Text): Part 1*. Edited by J.C.C. Mays. Princeton: Princeton University Press.

———. (2001b). *The Collected Works of Samuel Taylor Coleridge: Poetical Works I: Poems (Reading Text): Part 2*. Edited by J.C.C. Mays. Princeton, Princeton University Press.

Coleridge, Sara. (1873). *Memoirs and Letters of Sara Coleridge*. Edited by Edith Coleridge, 2nd ed., vol. 1. London: Henry S. King.

Darwin, Erasmus. (1791). *The Botanic Garden*. London: J. Johnson.

De Quincey, Thomas. (2003). "The English Mail-Coach, or the Glory of Motion." In Robert Morrison, ed., *The Works of Thomas De Quincey, Vol. 16: Articles from* Tait's Edinburgh Magazine, MacPhail's Edinburgh Ecclesiastical Journal, *the* Glasgow Athenaeum Album, *the* North British Review, *and* Blackwood's Edinburgh Magazine, *1847–1849*. London: Pickering & Chatto, pp. 408–28.

Holmes, Richard. (2008). *The Age of Wonder: How the Romantic Generation Discovered the Beauty and Terror of Science*. New York: Pantheon Books.

Kemble, Frances Ann. (1878). *Record of a Girlhood*. Vol. 2. London: Richard Bentley and Son.

Malm, Andreas. (2016). *Fossil Capital: The Rise of Steam Power and the Roots of Global Warming*. London: Verso.

Miller, David Philip. (2009). *James Watt, Chemist*. London: Pickering & Chatto.

Parker, Henry Meredith. (1834). "Thoughts on Flying." In D.L. Richardson, ed., *The Calcutta Literary Gazette or Journal of Belles Lettres, Science, and the Arts*, new series 1 (January 4), 3–4.

Peacock, Thomas Love. (1831). *Crotchet Castle*. London: T. Hookham.

Poe, Edgar Allan. (1902). "The Unparalleled Adventure of One Hans Pfaall." In Charles F. Richardson, ed., *The Complete Works of Edgar Allan Poe*, vol. 2, *Tales*. New York: The Knickerbocker Press, pp. 50–130.

Rabinbach, Anson. (1990). *The Human Motor: Energy, Fatigue, and the Origins of Modernity*. New York: Basic Books

Ricardo, David. (1951). *The Works and Correspondence of David Ricardo: Volume 1, On the Principles of Political Economy and Taxation*. Edited by Pierro Sraffa and M.H. Dobb. Cambridge: Cambridge University Press.

Russell, John Scott. (1846). *A Treatise on the Steam-Engine from the Seventh Edition of the Encyclopædia Britannica*. Edinburgh: Adam and Charles Black.

Schatzberg, Eric. (2018). *Technology: Critical History of a Concept*. Chicago: University of Chicago Press.

Scott, Heidi C.M. (2018). *Fuel: An Ecocritical History*. London: Bloomsbury Academic.

Shelley, Percy Bysshe. (1964). *The Letters of Percy Bysshe Shelley, Volume 2: Shelley in Italy*. Edited by Frederick L. Jones. Oxford: Clarendon Press.

Simpson, David. (2009). *Wordsworth, Commodification and Social Concern: The Poetics of Modernity*. Cambridge: Cambridge University Press.

Smith, Adam. (1993). *An Inquiry into the Nature and Causes of the Wealth of Nations: A Selected Edition*. Edited by Kathryn Sutherland. Oxford: Oxford University Press.

Somerville, Mary. (1834). *On the Connection of the Physical Sciences*. London: John Murray.

Speitz, Michele. (2024 [forthcoming]). *The Romantic Sublime and Representations of Technology*. Liverpool: University of Liverpool Press.

"Steam Engine to Water the Streets of Calcutta." (1823). *The Asiatic Journal*, 15(89), 524.

"Steam; and the Strides It Is Making." (1822). *The Times*, September 19, 3B.

Stuart, Robert. (1824). *A Descriptive History of the Steam Engine*. 2nd ed. London: Knight and Lacey.

Tresch, John. (2012). *The Romantic Machine: Utopian Science and Technology after Napoleon*. Chicago: University of Chicago Press.

Underwood, Ted. (2005). *The Work of the Sun: Literature, Science, and Economy, 1760–1860*. New York: Palgrave Macmillan.

Wood, William. (1833). *A Series of Twenty-Eight Panoramic Views of Calcutta, Extending from Chandpaul Ghaut to the End of Chowringhee Road, together with the Hospital, the Two Bridges, and the Fort*. London: William Wood.

Wordsworth, William. (2004). *Sonnet Series and Itinerary Poems, 1820–45*. Edited by Geoffrey Jackson. Ithaca: Cornell University Press.

———. (2007). *The Excursion*. Edited by Sally Bushell, James A. Butler, and Michael C. Jaye. Ithaca: Cornell University Press.

CHAPTER 3

Diffusions, Relocations, and the Permeative Process of Coalescence

Abstract Accounts of the first stationary engine in Bengal, at the Baptist Mission Press, and the public launch of the first steamboat to navigate the Hugli River show how the depiction of steam as sublime firmly fixed technology as a power that would diffuse European civilization. Against this ideology, the chapter positions accounts of the engine built by Golak Chandra, a blacksmith who took the Mission's engine as a prototype. Displayed at an exhibition of the Agricultural and Horticultural Society of India, this engine and the context of its exhibition capture tensions within the period's cultures of science and interracial sociability between what Kapil Raj has called "diffusion" and "relocation." The chapter concludes with an analysis of the Society's membership, a politically diverse, "cosmopolitan" group of Europeans and elite Bengalis including Dwarkanath Tagore and other figures intimately connected to Goodeve and Parker. Before turning to their futurist fictions in the next two chapters, I propose that Goodeve should be associated with what Uday Singh Mehta has called the "cosmopolitanism of reason," which was rooted in utilitarianism, while Parker embodied Mehta's "cosmopolitanism of sentiment," a Burkean perspective informing Parker's vision of future home rule for India and his skeptical depiction of maximalist global capitalism in "The Junction of the Oceans."

D. E. White, *Romanticism, Liberal Imperialism, and Technology in Early British India*, https://doi.org/10.1007/978-3-031-60705-9_3

Keywords Steam power in early British India · Golak Chandra · Agricultural and Horticultural Society of India · Kapil Raj · Liberal imperialism · Uday Singh Mehta

From March 27, 1820, when "The 'machine of fire,' as they called it, brought crowds of natives to the mission, whose curiosity tried the patience of the engineman imported to work it" (G. Smith, 1885, p. 245), to November 4, 1822, when *The Bengal Hurkaru* celebrated the agreeable chastity and proportion of "the beautiful steam engine erected at Chaundpaul Ghaut" ("Steam Engine," 1823, p. 524), to August 9, 1823, when the Diana Packet "left Chandpaul Ghaut at 11 a.m. of Saturday" ("The Diana," 1824, p. 280), Bengal was becoming the epicenter of an imperial imagination in which steam in particular was firmly fixed as an engine that would diffuse European civilization.

Bearing a party including Colonel Jacob Krefting, Governor of Serampore, and his suite, the Diana steamed to Chinsurah against a strong tide, making the journey in six to seven hours and then returning by way of Serampore, where an "elegant entertainment" had been "prepared for the occasion" ("The Diana," 1824, p. 280), arriving back in Calcutta on the following morning. Here is the account reproduced by *The Asiatic Journal* in London from *The Calcutta Journal*:

> As the vessel passed up, the banks of the river were crowded with natives, gazing with stupid wonder on this novel scene. To behold a vessel thus stemming a furious tide, without the aid of oar or sail, and sending forth from a black column, standing in the usual place of a mast, a volume of smoke, was indeed a sight well calculated not only to excite the curiosity, but to work on the superstitious fears of the natives; they gazed on it with silent amazement, or with loud expressions of astonishment, as the feelings of fear or curiosity predominated, utterly unable to divine the power by which the vessel was impelled with such velocity. Such was the effect of this specimen of the triumph of science over the elements, on some of the more ignorant natives, that several of them, it is said, actually leaped out of their boats into the river through fear. (p. 280)

All the terms of the Romantic sublime are there, yet the wonder is "stupid," the fears "superstitious," the amazement or astonishment, whether "silent" or "loud," indifferently a sign of utter incomprehension. For Burke and, especially, for Kant, what we call sublime does not reside in the object so described but rather in the mind that apprehends it. Kant is explicit that "the judgement upon the Sublime in nature needs culture (more than the judgement upon the Beautiful)": "That the mind be attuned to feel the sublime postulates a susceptibility of the mind for ideas ... In fact, without the development of moral ideas, that which we, prepared by culture, call sublime presents itself to the uneducated man merely as terrible" (Kant, 1951, p. 105). Unprepared by culture, the crowds of natives who gather to see the "machine of fire" in 1820 experience only curiosity, while in 1823 "the more ignorant natives" experience only terror. Having already noted that "Almost all the heathen temples were dark" (Burke, 1759, p. 100), Burke, in fact, denominated darkness and blackness themselves as intrinsically terrible objective properties, citing the example of a boy who "saw a black object," which "gave him great uneasiness," and then "some time after, upon accidentally seeing a negro woman, ... was struck with great horror at the sight" (p. 276). In accounts of steam as sublime, accordingly, it is the European viewer whose mind is elevated and "admitted ... into the Counsels of the Almighty by a consideration of his works" (p. 88). Thus elevated, the European mind is prepared, as *The Calcutta Journal* concludes its depiction of the Diana Packet, to "promote the cause of science and the arts, and add to the sum of human enjoyments" ("The Diana," 1824, p. 281). The native, on the contrary, like a dark, heathen temple, can be an object, not a subject, of the sublime. Whether by accident, design, or the sheer force of culture, the next article, following on the same page immediately below, recounts a "Suttee at Meerut," in which the writer rhetorically asks, "What feelings ... can these Hindoos have in common with us, who can thus calmly see their children or sisters put to death, and who can look on, not merely with indifference, but with delight" ("Suttee," 1824, p. 282). The language and associations were so pervasive that Mary Somerville surely did not think twice when, in *On the Connection of the Physical Sciences* (1834), she celebrated "the application of heat to the various branches of the mechanical and chemical arts" thus: "Armed by the expansion and condensation of fluids with a power equal to that of the lightning itself, conquering time and space, he ["man"] flies over plains, and ... like a magician, he raises from the gloomy and deep abyss of the mine, the spirit of light to dispel the midnight darkness" (Somerville, 1834, pp. 249–50).

The curious crowds that came to witness the Diana dispelling the darkness recall the contrast established in the account of the Serampore engine between collective and childlike native curiosity, which "tried the patience of the engineman," and the individualized and mature curiosity of "many a European who … came to study and to copy it." One Indian not only came to study and copy it, however, but then proceeded to build a working engine based on its design. We do not know if Golak Chandra, blacksmith of Titagarh, immediately opposite Serampore, crossed the river on March 27, 1820, to join the "crowds of natives" curious about the new machine. But he did cross the river to study it many times thereafter: at its meeting on January 9, 1828, the Agricultural and Horticultural Society of India resolved, "at the suggestion of the Rev. Dr. Carey, that permission be given to Goluk-chundra, a blacksmith of Titigur, to exhibit on Wednesday next, a Steam Engine made by himself without the aid of any European artist" (*Transactions*, 1838, p. 252).[1] Golak's engine was then displayed at the Society's annual exhibition on January 16 at Calcutta's Town Hall. In its report, the Society further emphasized that Golak designed and built the engine "without any assistance whatever from European artists upon the mode, of a large Steam Engine belonging to the missionaries at Serampore," for which the "ingenious blacksmith" was awarded a prize of 50 rupees (p. 257). A contemporaneous account in *The Calcutta Gazette* echoed the phrase "without any assistance whatever from European artists" while adding its own praise for the "native ingenuity" and "imitative skill" of the "ingenious Blacksmith" (Das Gupta, 1959, p. 273).

The story of Golak Chandra's engine, for Amitabha Ghosh, demolishes a myth, making the "plotting of Indian dependence as a function of technological incapability even in the era of steam a spurious fabrication" (Ghosh, 1993, p. 168). But something else is at stake too in the early nineteenth century. If we zoom out to the context in which this engine was exhibited in 1828, at an exhibition of the Agricultural and Horticultural Society, we find, according to the Society's Prospectus, written by the Baptist missionary William Carey on April 15, 1820, a mere three weeks after the demonstration of the Mission's new steam engine, that "It is peculiarly desirable that Native gentlemen should be eligible as members of the Society, because one of its chief objects will

---

[1] See Ghosh (1993) and Suvobrata Sarkar (2022).

be the improvement of their estates, and of the peasantry which reside thereon. They should therefore not only be eligible as members, but also as officers of the Society in precisely the same manner as Europeans" (*Transactions*, 1838, p. 220).[2] A further reason why it was peculiarly desirable to associate "Native Gentlemen of landed estates with Europeans who have studied this subject" was that "we should gradually impart to them more correct ideas of the value of landed property, of the possibility of improving it, and of the best methods of accomplishing so desirable an end" (p. 213). Such "improvements" would bring about "the gradual conquest of the indolence which in Asiatics is almost become a second nature,—and the introduction of habits of cleanliness ... in the place of squalid wretchedness, neglect, and confusion" (p. 214). Through the diffusion of western science to the native members of the Society, "industry and virtue" would replace "idleness and vice" (p. 214).

In this light, the insistence that Golak's ingenuity was a form of "imitative skill" and the emphasis here on the "improvement" of native estates (Britons being forbidden from owning land) reveal a tension in the period between what Kapil Raj has called the later historiographical perspectives of "diffusionism" and "relocationism" with respect to the creation and transmission of scientific knowledge. According to the dominant diffusionist model, science is developed in the West and then spread to the East as "the embodiment of basic values of truth and rationality, the motor of moral, social, and material progress, the marker of civilization itself" (Raj, 2007, p. 4). Knowledge and technologies are fundamentally stable in function and significance, adapting the new location to the universal values and progress they bear while remaining themselves unmodified by local environment or experience. The alternative "relocationist" perspective proposed by Raj, on the contrary, rooted in the New Imperial History's emphasis on asymmetric circulation and the mutual constitution of Britain and its empire (p. 7), suggests that "South Asia was not a space for the simple application of European knowledge" but rather "was an active, although unequal, participant in an emerging world order of knowledge ... [T]he contact zone was a site for the production of certified knowledges which would not have come into being but for

---

[2] It was subsequently resolved that the society would be administered by a President, two Vice-Presidents, two Secretaries, and a Treasurer; one of the Vice-Presidents and one of the Secretaries would always be a European and the other a native of India, with "The President to be always a European" (*Transactions*, 1838, p. 221).

the intercultural encounter between South Asian and European intellectual and material practices" (p. 13).[3] Undermining the equation between stable knowledge and the transfer of civilization from West to East, the production of science in the contact zone is always a form of hybridization bearing the marks of specific and unequal relationships of power.

On the one hand, then, Golak is a "mimic" engineer, copying a European imported prototype and demonstrating the native capacity, in the well-known language of Macaulay's Minute (1835), to become "English … in intellect" (Macaulay, 1972, p. 249). And the same goes for the 500 native members of the Agricultural and Horticultural Society over the remainder of the nineteenth century (G. Smith, 1885, p. 316), who, for Macaulay, would join the "interpreter" class—"a class of persons, Indian in blood and colour, but English in taste, in opinions, in morals, and in intellect"—prepared "to refine the vernacular dialects of the country, to enrich those dialects with terms of science borrowed from the Western nomenclature, and to render them by degrees fit vehicles for conveying knowledge to the great mass of the population" (Macaulay, 1972, p. 249).

On the other hand, the activities of the Society, and even at times the language of its *Transactions*, work against its diffusionist ideology: the "first endeavours" of the Society "would be directed to the obtaining of information upon the almost innumerable subjects which present themselves" (*Transactions*, 1838, p. 214), and native members would thus be essential not just to the civilizing process by which they would receive communicated European science but to the Society's own acquisition of a hybridized "stock of knowledge": "the methods employed to raise crops, and conduct the other parts of rural economy must so vary with soil, climate, and other local circumstances, as to make it impossible for any individual to be practically acquainted with them all. Too much praise can scarcely be given to local establishments whether public or private" (p. 214). Carey, Professor of Bengali, Sanskrit, and Marathi at the College of Fort William and as devout a practical botanist as a Particular Baptist, was far less invested in assimilation and anglicization than the

---

[3] Anna Winterbottom, among others, rightly cautions that the idea of "circulation" can be too broad to accommodate "a range of different encounters and methods of distribution" (Winterbottom, 2016, p. 20). More specifically, James Mulholland warns that the term can stand in for the very "structures and forces" that set it in motion; it is an effect, in other words, described as a cause (Mullholland, 2018, p. 374).

evangelical wing of the Church associated with Charles Grant, Zachary Macaulay, and the Clapham Sect, or than Alexander Duff of the Scottish Church Mission: above all, for Carey local circumstances (and languages) were always paramount. Here we see an example of what Miller calls "hybridization through the conjoining or commingling of foreign and local knowledge and skills" (Miller, 2017, p. 266).

The Society, in this respect, could not have been further in spirit from James Mill at his most Gradgrindian: heralding his own acknowledgment that the writer of *The History of British India* "has never been in India" (Mill, 1817, p. xii) as a qualification rather than a disadvantage, Mill asks what a man could possibly gain from direct experience, concluding, "He can treasure up the facts, which are presented to his senses; he can learn the facts which are recorded in such native books, as have not been translated; and he can ascertain facts by conversation with the natives, which have never yet been committed to writing. This he can do; and I am not aware that he can do any thing further" (p. xiv). As Majeed writes, "Mill argued that cultures and societies could be understood more comprehensively from a distance, when the sympathies of the commentator were not engaged in the detail of minute observations" (Majeed, 2005, p. 96). Because the process of translation for Mill is direct and transparent, facts can be transmitted unmodified through writing, and thus there is nothing further to know from local experience beyond facts that have not been so transmitted. Macaulay shares this understanding that the source language remains untouched while "enriching" the target language in his position that "the dialects commonly spoken among the natives of this part of India contain neither Literary nor scientific information, and are, moreover so poor and rude that, until they are enriched from some other quarter, it will not be easy to translate any valuable work into them" (Macaulay, 1972, p. 240). For Miller, on the contrary, "the rendering of Western scientific or technical knowledge in a local language ... will involve more than translation" (Miller, 2017, p. 266), and that surplus is the conjoining of knowledge and skills, beyond abstract theory, especially in what the period thought of as the "practical" arts, such as agriculture and engineering. At the meeting resolving the rules of the Society, it was decided that its transactions would be published in English and "two at least of the languages of India" (*Transactions*, 1838, p. 222), and here again the diffusionist and relocationist perspectives are audibly in tension: English enriches the languages of India, or the process of translation is one of the means by which the Society adds to its stock of knowledge by bringing English into contact with experience embedded in local circumstances. Finally, in a moment of simultaneous "hybridization of knowledge and of practices" in the contact zone (Miller, 2017, p. 265) and *reverse*-diffusionism, the Calcutta Agricultural and Horticultural Society founded in 1820 and shaped by intercultural encounter, translation, and local practices, then formed the model for the Royal Agricultural Society of England, founded in 1838 (G. Smith, 1885, p. 317).

Golak's engine was thus exhibited to a Society emblematic of a liberal ethos divided by conflicting impulses, both of which were in the service of empire, but of different visions of empire. And that conflict can be seen in the list of the Society's members published on July 1, 1828 (*Transactions*, 1838, pp. vii–viii), which was by no means unusual in bringing together a politically and ideologically diverse group of Europeans and elite Bengalis comfortable in one another's society, including the Baptist missionary and linguist Carey; the industrialist and entrepreneur Dwarkanath; editor of *The Reformer* (and cousin of Dwarkanath) Prasanna Kumar Tagore; future editor of *The Englishman* (and first cousin of Goodeve) William Cobb Hurry; Salt Agent and Board of Trade member (and close friend of both Dwarkanath and Parker) Trevor Plowden; conservative Hindu leader Radhakanta Deb and his fellow member of the Hindu College Board of Managers, the religiously orthodox but liberal-minded Ramkamal Sen; and the orientalist Horace Hayman Wilson, Hindu College professor and manager, Secretary of the General Committee of Public Instruction, and later chair of Sanskrit at Oxford and director of the Royal Asiatic Society.[4] After the Vellore Mutiny of 1806, Bentinck (then the

[4] The term "elite" needs to be qualified in the case of Dwarkanath and Prasanna Kumar, whose family was of the Pirali Brahmin caste, a "degraded Brahmin subcaste" (Kling, 1976, p. 11). Brahmins who dined with a Pirali would lose caste, and upper-caste sons would not marry the daughters of Pirali families. Pirali families, writes Kling, "met the challenge in one of two ways: at times they rejected Hindu tradition and adopted European manners and values; at other times they tried to prove their caste legitimacy by hyperorthodox behaviour" (p. 11). So although Dwarkanath rejected tradition and would sit with Europeans while they ate, equipping his villa in the English style with a billiard room and fitting it up "with statues and pictures, and Copley Fieldings, and Prouts, and French China, &c." (Eden, 1872, p. 215), during the years of his close association with his guru Rammohun he both attended services of the Brahmo Samaj (of which he nominally assumed the leadership after Rammohun's death in 1833) and continued to perform puja to his family idols, as he did for the remainder of his life (Kling, 1976, p. 22; Sartori, 2008, p. 90). Similarly, although Prasanna Kumar publically opposed idolatry in

Governor of Madras) isolated a source of Company misrule in the fact that Britons in India were "strangers in the land": "Europeans generally know little or nothing of the customs and manners of the Hindus ... We do not, we cannot, associate with the natives. We cannot see them in their houses and with their families ... [W]e are in fact strangers in the land" (qtd. in Rosselli, 1974, p. 146). As disparate as were the political, religious, and social perspectives of these individuals, they were deeply invested in overcoming estrangement by forging a "new British-Indian nationality" (Kling, 1976, p. 161), although—and this is the crux—the "British-Indian" nature of that new nationality varied dramatically.

One way to grasp this variety is by considering the different ways in which social associations and relationships among and between Europeans and Indians were understood and expressed. An 1839 article in *The Bombay Gazette* draws a stark distinction between Bombay and Calcutta along precisely these lines, proposing that in the latter "we do recognize something like community of feeling and a combined idiosyncrasy; societies, meetings, projections follow in quick succession, and a current of healthy sympathy and sentiment seems to pervade the monied mass": whereas Bombay is frozen in "the lonely icicled state of magnificence," in Calcutta "the thaw of social harmony has produced a permeative process of coalescence, which is spreading in every direction, and resolving into one community both Europeans and natives" (qtd. in Sartori, 2008, p. 76). As we will see, Dwarkanath was a major node in an extensive commercial, political, and social network, linking both Goodeve and Parker and making recognizable "something like a community of feeling" among "the monied mass," though one far more divided than the "permeative process of coalescence" would imply. Dwarkanath, Goodeve, and Parker, then, were part of a complex but by no means homogeneous "cosmopolitan" community that joined societies such as the Agricultural and Horticultural Society; advocated for scientific, technological, and

---

The Reformer, he nonetheless celebrated Durga Puja, for which he was criticized in the pages of Henry Louis Vivian Derozio's *The East Indian*. After Dwarkanath further rejected tradition by sailing to England in 1842, he suffered "the penalty of expulsion from his family circle" (Stocqueler, 1845, p. 41). In his assault on caste in *The Empire of the Middle Classes*, Parker wrote of an unnamed friend who is clearly Dwarkanath: "My friend was a man of noble benevolence in his own country, and courted by the very best society in this; but, in despite of his wealth, his charity, his influence, and accomplishments, the lowest and most beggarly Brahmin in his heart spat upon him, and scorned him with a scorn not to be described" (Parker, 1858, p. 18).

educational improvements; published poetry and fiction in newspapers, gazettes, and annuals; and shaped the contours of liberal imperialism. In order to see how consequential differences in the broadly liberal perspectives of Goodeve and Parker informed their divergent futurist fantasies in *The Bengal Annual*, we will need to trace some of their activities and friendships and to consider the different ways in which their interracial relationships unfolded.

Goodeve and Parker shared numerous social connections and were both strong advocates for the freedom of the press. They both wrote for David Lester Richardson's *The Bengal Annual*, as did Agricultural and Horticultural Society member W.C. Hurry, who was a close friend of Parker as well as Goodeve's first cousin.[5] In early 1838, both, along with Richardson and Prasanna Kumar, were stewards of a "dinner to be given to Sir Charles [Metcalfe] by those who appreciate his measure of freeing the Indian Press" ("Metcalfe Meeting," 1838, p. 94) in 1835. And both, as I have mentioned, were closely connected to Dwarkanath, who accompanied each on their final return voyage to England, Parker's in 1842 and Goodeve's in 1845.

But a more minute inspection of their activities and associations, as well as their fictions, reveals meaningful fissures within the pervasive liberal imperialism of the 1830s. Goodeve, I propose, should broadly be associated with what Uday Singh Mehta has called the "cosmopolitanism of reason" (Mehta, 1999, p. 20), which was rooted in the writings of James Mill, J.S. Mill, and Bentham, informed the liberal-utilitarian administration of Bentinck and Macaulay, and accorded with the interests of the free-traders dominated by the European nonofficial community, whose positions were promoted in the press by James Silk Buckingham, and with whom Rammohun Roy and Dwarkanath were generally aligned. For Goodeve, technology and scientific knowledge are utilitarian tools that diffuse civilization or, in the terms of his futurist fantasy, furnish modern

---

[5] W.C. Hurry was the son of Eliza Hurry (née Liddell) and Edmond Cobb Hurry, the brother of Henry Hurry Goodeve's mother, Elizabeth Goodeve (née Hurry). In 1842, Hurry purchased *The Englishman* newspaper from Stocqueler, under whom the paper had recently declared bankruptcy. (In 1833, Stocqueler had purchased *The John Bull* and quickly changed its name to *The Englishman* and its politics from Tory to Whig.) Under Hurry's editorial management (until 1856), the paper became the "most indulgent of friends, the most fostering of patrons" to the writers of Calcutta's literary scene and defended the interests of indigo planters (Chanda, 1987, pp. 169–72).

"luxuries" or "conveniences" and set in motion the global circulation of commodities.

Parker, on the other hand, whose nickname, it can't be stressed strongly enough, was "Proteus" (Stocqueler, 1873, p. 90), embodied what Mehta has called the "cosmopolitanism of sentiment," a fundamentally pragmatic and Burkean perspective that "acknowledges that the integrity of experience is tied to its locality and finitude ... By doing so it is congruent with the psychological aspects of experience, which always derive their meaning, their passionate and pained intensity, from within the bounded, even if porous, spheres of familial, national, or other narratives" (Mehta, 1999, p. 21).[6] In keeping with its less progressive sensibilities, this perspective informed what Blair Kling insightfully calls "that amorphous party of the loyal opposition" (Kling, 1976, p. 161), which harkened back to "the 'Romantic' generation in British-Indian history" (Stokes, 1959, p. 15) of Thomas Munro, John Malcolm, Charles Metcalfe, and (to a lesser extent) Mountstuart Elphinstone. Although "Largely in agreement with certain aspects of the Utilitarian viewpoint," as Eric Stokes writes, "to the spirit of utilitarianism they were as uncompromisingly hostile as Burke ... [T]hey countered the new spirit by an appeal to history and experience": "Politics to them were experiential in nature, necessarily near-sighted, and essentially limited in their achievement" (p. 23). This was the party of Parker, Hurry, Wilson, Richardson, Theodore Dickens, and, to a certain degree, Dwarkanath, who owed much of his influence to his protean capacity (shared by Rammohun) for fluency not just in the English language, as was widely remarked upon, but in European codes of friendship and business across various lines of

---

[6] Mehta's account of Burke's "cosmopolitanism of sentiment" dovetails with what Jennifer Pitts calls his "peculiar universalism" (Pitts, 2005, pp. 59–101). Sartori has critiqued Mehta's categories, historically and conceptually. Historically, Sartori suggests that the sympathetic interpretation of Burke does not account for the range of his positions on India. Conceptually, he argues that Mehta's opposition between the concreteness of experience and liberal abstraction, with the former serving to critique the latter, is untenable because concrete social forms are never separate from or outside of forms of social abstraction: "for example, the modern household in its relationship to the wage and to commodity consumption" (Sartori, 2008, pp. 26–27). I agree, especially with the historical critique, but I am using Mehta's valuable terms to a different end, not in order either to interpret Burke or to offer a critique but rather as powerful descriptive (and heuristic) categories that accurately capture the language and ways in which early-nineteenth-century liberals themselves were unevenly and inconsistently placing themselves in relation to a range of ideas and feelings (personal, political, imperial) and to one another.

interest. Technology in Parker's "The Junction of the Oceans" is not a tool but rather an extension of humanity's power, of its delusive mastery of the earth, which mankind has subordinated to a single purpose—global commerce. But humanized power, in Parker's future, remains subject to the same invisible and higher power of the divine plan for both earth and humanity, a plan we can sometimes see written clearly, in catastrophes, and sometimes only guess at, tentatively, through our own sentiments and failings, and through the signs that God inscribes on and in our world.

Both sides of liberal imperialism ardently promoted steam power in India, the "great engine" of India's "moral improvement," in Bentinck's words (*Select Committee*, 1837, p. 128), but, as we will see when we now turn to Goodeve's and Parker's lives and fictions, with very different inflections.

References

Burke, Edmund. (1759). *A Philosophical Enquiry into the Origin of Our Ideas of the Sublime and Beautiful*. 2nd ed. London: R. and J. Dodsley.

Chanda, Mrinal Kanti. (1987). *History of the English Press in Bengal 1780 to 1857*. Calcutta: K.P. Bagchi.

Das Gupta, Anil Chandra, ed. (1959). *The Days of John Company: Selections from Calcutta Gazette 1824–1832*. Calcutta: Government Printing.

"The Diana Steam Packet." (1824). *Asiatic Journal*, 7(99), 280–81.

Eden, Emily. (1872). *Letters from India*. Edited by Eleanor Eden, vol. 1. London: Richard Bentley and Son.

Ghosh, Amitabha. (1993). "Golak Chandra: India's Pioneer Innovator Technician." *Indian Journal of History of Science*, 28(2), 167–78.

Kant, Immanuel. (1951). *Critique of Judgment*. Translated by J.H. Bernard. New York: Hafner Press.

Kling, Blair B. (1976). *Partner in Empire: Dwarkanath Tagore and the Age of Enterprise in Eastern India*. Berkeley: University of California Press.

Macaulay, Thomas B. (1972). "Minute on Indian Education." In John Clive and Thomas Pinney, eds., *Thomas Babington Macaulay: Selected Writings*. Chicago: University of Chicago Press, pp. 237–51.

Majeed, Javed. (2005). "James Mill's *The History of British India*: The Question of Utilitarianism and Empire." In Bart Schultz and Georgios Varouxakis, eds., *Utilitarianism and Empire*. Lanham, MD: Lexington Books, pp. 93–105.

Mehta, Uday Singh. (1999). *Liberalism and Empire: A Study in Nineteenth-Century British Liberal Thought*. Chicago: University of Chicago Press.

"Metcalfe Meeting." (1838). *The Calcutta Monthly Journal*, 39, 94.

Mill, James. (1817). *The History of British India*. Vol. 1. London: Baldwin, Cradock, and Joy.

Miller, David Philip. (2017). "The 'Relocation' of Technology Between East and West: Stationary Steam Engines and Steamboats in India in the Early Nineteenth Century." In Sundar Sarukkai, ed., *Science and Narratives of Nature: East and West*. London: Routledge, pp. 261–303.

Mullholland, James. (2018). "An Indian It-Narrative and the Problem of Circulation: Reconsidering a Useful Concept for Literary Study." *Modern Language Quarterly*, 79(4), 373–96.

Parker, Henry Meredith. (1858). *The Empire of the Middle Classes. Being Nos. 1 and 2 of Short Sermons on Indian Texts*. London: W. Thacker.

Pitts, Jennifer. (2005). *A Turn to Empire: The Rise of Imperial Liberalism in Britain and France*. Princeton: Princeton University Press.

Raj, Kapil. (2007). *Relocating Modern Science: Circulation and the Construction of Knowledge in South Asia and Europe, 1650–1900*. London: Palgrave Macmillan.

Rosselli, John. (1974). *Lord William Bentinck: The Making of a Liberal Imperialist, 1774–1839*. Berkeley: University of California Press.

Sarkar, Suvobrata. (2022). "The Small Voices of History: Subaltern Technologists of Colonial Bengal." In Kaustubh Mani Sengupta and Tista Das, eds., *Rethinking the Local in Indian History: Perspectives from Southern Bengal*. Abingdon: Routledge, pp. 89–105.

Sartori, Andrew. (2008). *Bengal in Global Concept History: Culturalism in the Age of Capital*. Chicago: University of Chicago Press.

Select Committee on Steam Communication with India. (1837). London: House of Commons.

Smith, George. (1885). *The Life of William Carey, D.D. Shoemaker and Missionary. Professor of Sanskrit, Bengali, and Marathi in the College of Fort William, Calcutta*. London: John Murray.

Somerville, Mary. (1834). *On the Connection of the Physical Sciences*. London: John Murray.

"Steam Engine to Water the Streets of Calcutta." (1823). *The Asiatic Journal*, 15(89), 524.

Stocqueler, J.H. (1845). *The Hand-Book of India, a Guide to the Stranger and the Traveller, and a Companion to the Resident*. 2nd ed. London: Wm. H. Allen.

———. (1873). *The Memoirs of a Journalist*. Bombay: The Office of the Times of India.

Stokes, Eric. (1959). *The English Utilitarians and India*. Oxford: Clarendon Press.

"Suttee at Meerut." (1824). *Asiatic Journal*, 7(99), 281–82.

Transactions of the Agricultural and Horticultural Society of India. (1838). Vol. 1. Calcutta: Baptist Mission Press.

Winterbottom, Anna. (2016). *Hybrid Knowledge in the Early East India Company World.* Basingstoke: Palgrave Macmillan.

CHAPTER 4

Henry Hurry Goodeve and Dominion over the Wants of the Universe

Abstract Two futurist fantasies serve as segues from utilitarian and Romantic accounts of steam in the long eighteenth century to a reading of Henry Hurry Goodeve's "1980." First, Félix Bodin's *Le roman de l'avenir* [*The Novel of the Future*] (1834) presents a global allegory set in the late twentieth century, pitting the Benthamite "Universal Congress" against the "Poetic or Anti-Prosaic Association," which is allied to the "despots" of Asia. Next, "A Family Conversation," published in Calcutta in the same year, records a conversation among an Indian family in 1935, "a liberal age" of "enlightened utility." The chapter then turns to Goodeve's career as a Company medical doctor. His energetic support for the education of elite Indians and supreme confidence in the civilizing mission of their advancement were not just compatible but, indeed, flip sides of the cosmopolitanism of reason. This is the frame in which I read "1980," in which technology is a utilitarian tool that has transformed the "Indian Republic" into an anglicized techno-wonderland and liberal imperial power. The optimistic future imagined by Goodeve accords with visions of Indian nationalism in which independence will be the natural result of diffusion, the circulation of commodities, modernization, and assimilation, not, as in Kylas Chunder Dutt's "A Journal of 48 Hours of the Year 1945" (1835), the inevitable aftermath of revolt.

Keywords Henry Hurry Goodeve · "1980" · Félix Bodin · "A Family Conversation"

© The Author(s), under exclusive license to Springer Nature
Switzerland AG 2024
D. E. White, *Romanticism, Liberal Imperialism, and Technology in
Early British India*, https://doi.org/10.1007/978-3-031-60705-9_4

Bodin's *Le roman de l'avenir* (1834) can serve as a paradigm of the futurist imagination in the mid-1830s, introducing the key terms that "A Family Conversation" (1834) and Goodeve's "1980" (1835) adapt to the conditions of British India in those years. Equal parts utopian and dystopian allegory prompted by utilitarian fantasies and sentimental fears, Bodin's late-twentieth-century world, accordingly, is divided in two. On the one side is the "congrés universel" ["universal congress"], which assembles in the Central American Republic of Benthamia (Bodin, 1834, p. 145; 2008, p. 95), and its military wing, the "*association civilisatrice*" ["association for civilization"] (p. 189; p. 116).[1] The other side is subdivided between the "polygames, despotes et consorts" ["polygamists, despots and their associates"] of Asia, on the one hand, and the European "association poétique ou *anti-prosaïque*" ["Poetic or Anti-Prosaic Association"], on the other hand, which the Universal Congress accuses of carrying out "procédés réellement anti-humanitaires" ["truly anti-humanitarian activities"] (p. 189; p. 116), a catalog of which will follow below. Steam-powered "aérostats" bring the delegates of the Universal Congress to Benthamia, although the initial suggestion was to hold the assembly "dans les airs, comme on le fit, il y a plusieurs années, à Calcutta" ["in mid-air, as had been done a few years before in Calcutta"] (p. 150; p. 97). Irony pervades Bodin's portrayals of both perspectives. Introducing the Universal Congress, Bodin explains, "Le mouvement est la condition de la vie. Puis qu'il en est ainsi, jamais on n'a tant vécu que de notre temps" ["Movement is the condition of life. Since that is the case, people have never lived as much as they do in our time"] (p. 141; p. 93). "La faculté locomotive" ["The locomotive faculty"] has effected "le croisement des races, le mélange des nations … de plus en plus profondément" ["the interbreeding of races and the mixing of nations … more and more profoundly"], and "le mot de *nationalité* commence à n'avoir plus qu'une vague signification" ["the word 'nationality' is beginning to lose any but a vague significance"] (p. 142; p. 93). Through the utilitarian policies of the Universal Congress, "la civilisation européenne a fait des prodiges de colonisation dans les deux hémisphères depuis près de

---

[1] Throughout this paragraph, after the French from Bodin (1834) I provide the English from Brian Stapleford's translation, listed in the references as Bodin (2008). Parenthetically, I first provide the pagination from Bodin (1834) followed by the pagination from Bodin (2008).

deux siècles" ["European civilization has accomplished prodigies of colonization in the two hemispheres in the last two centuries"], achieving "l'abolition de l'esclavage politique et domestique" ["the abolition of political and domestic slavery"] (pp. 292–93; pp. 162–63). But liberal utopia is not without its pitfalls:

> D'autre part, la mobilité, l'inquiétude, européennes, se sont donné ample carriére pour exploiter et peut-être un peu pour tourmenter ce pauvre globe. La vie sédentaire, intérieure at tranquille; les affections de la famille, l'amour du foyer domestique, de ce que l'ancienne Angleterre appelait le *home*; l'habitude douce et triste à la fois de voir les même arbres, les même rochers, le même clocher, disons plus, les même tombeaux, les tombeaux de nos péres, de nos proches, de nos amis: tout cela semblait près de se perdre, si les sentimens qui tiennent intimement à la nature de l'homme n'étaient pas indestructibles. (pp. 142–43)

> [On the other hand, European mobility and anxiety have been given ample scope to exploit and perhaps even to torment the poor globe. Sedentary, internal and tranquil life, family affections, the love of the domestic hearth—which old England calls "home"—the simultaneously gentle and sad habituation to seeing the same trees, the same rocks, the same belfry, not to mention the same tombs, the tombs of our fathers, our kin, our friends, would all seem near to being lost if the most stubborn sentiments of human nature were not indestructible. (p. 94)]

The pendulum, accordingly, has been swinging back toward sentiment and an "esprit de fixité" [a "spirit of fixity"]: "Un point d'arrêt s'est trouvé; et non-seulement la fureur locomotive s'est calmée, mais encore le repos, l'esprit de fixité, ont repris grande faveur" ["A stopping point has been reached, and not only had the locomotive fury abated, but the spirit of fixity has regained considerable favor"] (pp. 143–44; p. 94). Sounding much like Wordsworth at his most Burkean, in "Musings near Aquapendente April, 1837"—"By gross Utilities enslaved we need / More of ennobling impulse from the past, / If to the future aught of good must come" (Wordsworth, 2004, p. 756)—the Anti-Prosaic Association seeks to save "de curieuses ruines d'églises, de mosquees, de pagodes, de châteaux, d'abbayes, et pour conserver entiers les anciens monuments dont l'industrie était sur le point de s'emparer pour les arranger à son usage ou pour les détruire" ["curious ruins of churches, mosques, pagodas, chateaux and abbeys, and to conserve entire ancient monuments of which industry was on the point of taking possession, in order to convert them to its own use or demolish them"], to restore "tant de beaux modèles des architectures lombarde, gothique, sarrasine, hindous, etc." ["so many beautiful examples of Lombard, Gothic, Saracen and Hindu and many other achitectures"], to rescue "des dialectes qui avaient été des langues de peuples célèbres ... de l'entier abandon où ils sont tombés" ["dialects that had been the languages of famous peoples ... from the total disuse into which they had fallen"], to reassemble Stonehenge, and to preserve "vastes et tristes espaces couverts de la sombre et poétique bruyère" ["vast and dismal spaces covered by dark and poetic heather"], many of which have been "labouré, semé et moissonné par l'inévitable et imperturbable vapeur" ["ploughed, sown and harvested every year by the inevitable and imperturbable power of steam"] (Bodin, 1834, pp. 299–302; 2008, pp. 166–67). The English section of the Anti-Prosaic Association then reports on "l'état des mines de charbon de terre dans les îles britanniques" ["the state of coal mining in the British Isles"], offering "la perspective consolante d'un futur épuisement de cet odieux aliment de l'industrie mécanique, de ce puissant agent d'une civilisation triste, uniforme, monotone, destructive de toute vie poétique" ["the consoling perspective of a future exhaustion of that odious aliment of mechanical industry, that powerful agent of a dismal, uniform and monotonous civilization destructive of all poetic life"] (p. 303; p. 168).

Another futurist fantasy of 1834 that also dramatizes the tensions between reason and sentiment, to borrow the terms of Hazlitt's "The New School of Reform; A Dialogue between a Rationalist and a Sentimentalist" (1826), appeared in Richardson's *The Calcutta Literary Gazette*. "A Family Conversation" is set in Calcutta, 1935, "a liberal age" of "enlightened utility" in which there is "No poetical zig-zag" ("A Family," 1834, pp. 354–55).[2] The conversation takes place in "a spacious room, marble paved," with "two thermantidotes winnowing the air to coolness—an open door discloses a conservatory, odorous with Australian and American exotics" (p. 353).[3] The family comprises the utilitarian parents, Raj Kishen Roy, "A Hindoo Gentleman" who spends "an

---

[2] On "zig-zag" as a code for a kind of imperial light reading that embraced local experience and rejected the dominant literary ethos of exilic nostalgia, see White (2016).

[3] A thermantidote was a large cooling device that used hand-operated fans to draw air from outside over wet khus-khus grass. Fanny Parkes provides an excellent description (Parkes, 2001, pp. 149–50). See below, Chapter 5, note 8.

enormous outlay in improvements and machinery" for his "laboratory," which contains "ten thousand rupees' worth of alembics, cylenders [*sic*], retorts, &c." (p. 355), and his mercenary wife Misree; their frivolous, sentimental daughters Solah, who refuses to attend "anatomical experiments" (p. 354) and is in love with Alcibiades Muggins, impoverished younger son of a now wealthy family and grandson of "a soap-boiler" (p. 353), and Goolaubee, who sighs over "poor, dear Werter!—cold, stupid Charlotte!" (p. 354) and longs to visit England so she might "pluck a bough of the old bay-tree that grows over Byron's tomb at Newstead" (p. 355); their studious son, Erasmus Juggoo, who has just returned from a scientific demonstration at the Allypore amphitheater, where he received a black eye from the galvanized tail of an "electrified rhinoceros" (p. 354); and their diplomatic servant, Pauch Kouree, an "enfant trouve" (p. 356). Although Misree is full of praise for Muggins' "taste in lace, balloon patterns, and utilitarian novels," she is keen for Solah to marry the wealthy but "odious Sir Rammohun Buchanan," who has a glass eye and a cork leg, and is "on the wrong side of fifty" (p. 353). (In the end, the elder Muggins brother breaks his neck while out hunting "with the Hobarts-town pack" [p. 356], a happy resolution for Alcibiades and Solah.) Sir Rammohun Buchanan's name cleverly mashes together Rammohun Roy (thus the family's surname), India's "first liberal" (Bayly, 2004, p. 293), advocate of Western education, and monotheistic leader of the reformist Hindu community until his death in 1833, when Dwarkanath succeeded to the role, and Rammohun's friend Buckingham, slashing editor of the radical *The Calcutta Journal*, strident free-trader, and fierce critic of the Company both before and after his expulsion from India in 1823.[4] In this satirical depiction of a mechanized, liberal future, global capitalism blends the exotics of Australia and America, European and Asian fabrics and fashions, and both class and race into an undifferentiated, thermantidote-cooled mass, and it is as impossible to tell where England ends and anglicized, deracinated India begins as it is to separate Rammohun from Buckingham. Whereas Bodin's satire skewers both the Universal Congress and the Poetic or Anti-Prosaic Association, locating the extremes of modernity and tradition, of liberalism

---

[4] For important early discussions of Rammohun's liberalism, see De (1975), Sumit Sarkar (1975), and Sen (1975), and more recently Bayly (2004, pp. 50–60), Chatterjee (2012, pp. 134–58), Chaudhuri (2018), Koditschek (2011, pp. 90–98), Sartori (2008, pp. 77–89), White (2013, pp. 23–28, 110–20), and Zastoupil (2010).

and despotism, reason and sentiment, prose and poetry, in Asia in general, where polygamy and slavery persist, and in Calcutta in particular, where the assembly of the Universal Congress had previously met in mid-air on steam-powered aerostats, "A Family Conversation" suggests in terms very similar to Goodeve's in "1980" that the clash between reason and sentiment masks an unchanging and essentially English human nature, which, through the diffusion of technology and the abridgment of time and space, will become universal, eliding distinctions of nation, race, and class, but emphatically not of wealth, for capital makes this world go round: the European grandson of a soap-boiler gets the Hindoo Gentleman's daughter, but only after succeeding to an inheritance.

Trained in London and Edinburgh, Goodeve (Fig. 4.1) received his MD at age 21 in 1828, establishing himself back in London as an authority on midwifery. In April 1831, having been appointed Assistant Surgeon on the Bengal Establishment, he sailed for India, where he was stationed at Rampur.[5] In 1833, while on medical leave in Calcutta (having received a severe gunshot wound to the face during a tiger hunt), he wrote a letter to the General Committee of Public Instruction critical of the Native Medical Institution, which had been established in 1822 and offered training in Bengali in both Western and Ayurvedic medicine. Goodeve's first recommendation was that "The knowledge of the English language is an indispensable qualification for admission and all instruction should be carried out in English"; second, "Practical anatomy should be the basis for the surgical instruction," he advised, suggesting that "The prejudices of some Hindus are not insurmountable"; and third, "a better class of young men," which is to say upper-caste Hindus, should be admitted, paid a higher salary, and classified as "sub-assistant surgeons" (Whitfield, 2022, p. 7). The liberal-utilitarian agenda is clear, from the advocacy for training in English rather than the vernacular to the view that Hindu customs were superstitions to be surmounted and the promotion of elite Indians to positions of authority within the Company's service. Michael Whitfield writes that this letter "supported Governor-general William Bentinck's reformist Anglicist views and resulted in Goodeve being appointed assistant to Dr MJ Bramley as Superintendent of a new

---

[5] Upon his arrival, his wife's uncle, William Cameron, reportedly asked him what kind of appointment he was seeking: "'I should like to get a staff-appointment in Calcutta,' replied Goodeve. 'Young man,' said Cameron, 'would you like a slice of the moon?'" (Beddoe, 1910, p. 166).

Fig. 4.1 "H.H. Goodeve Esqre. M.D.," by Colesworthey Grant, from *The India Journal of Medical and Physical Science* (1837, facing p. 429); from the British Library Archive shelfmark ST 526[6]

English-speaking Calcutta Medical College that was established in 1835" (p. 8). That appointment inaugurated his association with Dwarkanath,

---

[6] This lithographic sketch of Goodeve from Grant (1837) was also included in Grant ([1850?]) as Plate 7, but the plate is missing from the copy at the British Library.

who not only donated three annual sums of 2,000 rupees for distribution in prizes (Mittra, 1870, p. 27) but was instrumental in supporting dissection: Kissory Chand Mittra reports that "the constant presence of Dwarkanath in the dissecting-room did much to remove" the "general repugnance of the Hindus to dissection" (p. 28), and Goodeve would later describe the successful introduction of practical anatomy as "The most important blow which has yet been struck at the root of native prejudices and superstition" (Goodeve, 1847, p. 189). In the characteristically diffusionist language of an 1899 account, "One of the strongest prejudices of the Hindus has thus been overcome, and the first and most important step of a rational system of medicine in the East had been accomplished" (qtd. in Whitfield, 2010, p. 11). In 1845, Goodeve left India accompanied by Dwarkanath along with four medical students, Suraj Coomar Chuckerbutty, Bholanath Bose, Dwarkanath Bose, and Gopal Chundra Seal. With the financial backing of Dwarkanath, who would die in England in the following year, they studied at University College, which had been established in 1826 as a nonconformist and utilitarian alternative to hidebound and exclusionary Oxford and Cambridge. On August 5, 1846, Goodeve and the four students attended Dwarkanath's funeral in Kensal Green. And in 1847, Goodeve reported on their progress in *The Lancet*, writing, "To the medical profession it should ... be a subject of much interest, and some little pride, to behold our art, as on so many previous occasions, once again employed as the instrument of civilization and the means of introducing the lights of learning and refinement amongst nations plunged in comparative darkness" (Goodeve, 1847, pp. 189–90).

Mary Ellis Gibson suggests that Goodeve "seems to have been, for his time and station, remarkably unprejudiced," proposing that his language here may be "a defensive concession to his critics" (Gibson, 2019, pp. 80–81), but it is important to consider the extent to which energetic support for the education and advancement of elite Indians, on the one hand, and supreme confidence in the civilizing mission of such education and advancement, on the other, were not just perfectly compatible but, indeed, flip sides of the liberal "cosmopolitanism of reason." Thus Theodore Koditschek writes of James Mill, "So far from being a racist, he was a steadfast believer in human equality who repudiated all notions of biological inferiority ... With the benefit of proper education and training, he insisted, all people could achieve the same high level of civilization and morality. Yet this offer of political equality was made contingent on a refusal of recognition to Hindu culture. Mill's harsh indictment of Indians was thus intended to help them—to liberate them from the chains of their repellent history" (Koditschek, 2011, pp. 83–84). In the education of four Hindus in England, Goodeve concludes that medical practitioners have "become pioneers of civilization, and our profession has assumed an honourable prominence in the march of improvement" (Goodeve, 1847, p. 190). One of the four, Suraj Coomar, would convert to Christianity in 1848 and adopt Goodeve's name, becoming Suraj Coomar Goodeve Chuckerbutty (Whitfield, 2022, p. 12).

In Goodeve's "1980," Calcutta has become "the London of the East" (Goodeve, 1835, p. 150), and the "Indian Republic," now liberated from the chains of its religion and culture, is a techno-wonderland and liberal imperial power with colonies in Australia.[7] The plot relates the thwarted love between Charles Holkar, "a descendant of the ancient Mahratta chief of that name," who attended the Calcutta University and then made the grand tour not just of Europe but of "*the world*," and Emmeline Nursing, daughter of Sir Thomas Nursing, the Bengal Ambassador, "who was himself of pure European extraction" (p. 151). Sir Thomas refuses his consent on the basis not of Charles' race—on the contrary, "the blood of all the Holkars was not compensation enough for the want of that more gross element" (p. 159)—but rather his relative poverty. The plot becomes intricate, but having met and lost the girl, the boy gets her back by rescuing the Nursings, along with Sir Thomas' sister-in-law, Mrs. Seebchunder, "member for Hazerabagh" in the newly reformed Indian parliament (p. 161), from the Calcutta prison, to which they had been sentenced through the machinations of the powerful and wealthy Sir Chytun Singh, district judge of Tibet and rejected suitor of Emmeline. Singh had deceived a servant, Nooran, into placing spurious documents among Mrs. Seebchunder's private papers implicating her as a secret supporter of India's rival power, the State of New Holland. The climax involves a dramatic prison break by steam balloon followed by a cinematic balloon chase scene, which features some very fine nonsense: pursued by a larger and faster government steamer, our heroes set their course for the limited monarchy of China, still hoping to outrun their pursuers, "for in some respects their balloon was better constructed than the

---

[7] "1980" was the first of two contributions by Goodeve to *The Bengal Annual*: "The Doctor's Tale" followed in *The Bengal Annual* for 1836. For a more detailed summary of the plot of "1980," see Sarkar (2023, pp. 45–46).

enemy's, being propelled by the combustion of the quadriginto-nitrate of Rhodium, instead of the more clumsy fashion of the oxy-hydrogen blow-pipes, and pounded diamonds" (p. 177).[8] The party escapes to China, in whose service Charles departs on an expedition against the Esquimaux in which, having used boiling water to capture many of the ice forts but failed to breach the ice walls of the capital, he is assisted by a soldier who proves to be a cross-dressed Nooran. She tells him of a secret passageway by which he takes the citadel. Having deduced Nooran's unwitting role in Singh's plot against Mrs. Seebchunder, Charles rejoins the party, which returns to Calcutta, where a new trial is held, Nooran gives her evidence, the sentence is reversed, Chytun Singh is tried and executed, and, now that Charles has demonstrated his disinterested devotion and been enriched by the grateful government of Pekin, the happy couple is united.

The tale opens on the Himalaya steam mail or "Vaporo" (p. 148), "rolling swiftly along the polished trams of the great high rail road to Calcutta. The vehicle was one of the most elegant of modern improvements, fitted up with air-cooling machines, fountains of iced soda-water, and every other convenience" (p. 147). Recalling Baillie's steam vessel, "Freighted with passengers of every sort, / ... a floating fair" (Baillie, 1823, p. 259), the party on board "was of both sexes, and of all colors and ranks, for the republican spirit of the time allowed of no distinction: he that paid his fare was as good as his neighbour, so that generals, and buneahs, dowagers, and market women, sat together cheek by jowl, in amicable confusion" (Goodeve, 1835, p. 147). On the train Charles converses with "an elderly Brahmin ... dressed in black," who turns out to be a parson. To this Brahmin parson the Engineer of the Vaporo recommends "the Rhinoceros, decidedly the best house in town," where he will find "the best beef-steak, and iced claret in all India, and the house is famous for its refrigerating beds" (p. 148). The Brahmin replies, "talking of beef, commend me to that glorious sirloin we had at Benares,—egad! ... upon my word, our worthy Hindoo fathers, peace to their souls! knew

---

[8] Charles and Emmeline originally meet on a "party of pleasure" to "the Snow King's palace," an enormous grotto of ice in the arctic: "Steam ships fitted for the purpose ran backwards and forwards constantly in the season," and voyagers are protected from sea-sickness by a "patent anti-emetico-stomatic" and from the cold by "anti-frigorific cloaks" (Goodeve, 1835, p. 152).

not what they lost, when they in their blindness denied their stomach such luxuries" (p. 149).

It is worth pausing here to consider the resonance of the Brahmin's recollection of that glorious sirloin on a steam train in the future imagined from the perspective of 1835, when the memory of two items of news from 1831 would have still been fairly fresh. The first, a scandal, was commemorated in Parker's poem "A Bengal Pastoral" (1831), with which Goodeve's "1980" may here be at least indirectly in conversation.[9] Students of Henry Derozio at the progressive Hindu College—who came to be known as "Young Bengal" and were famously described as "cutting their way through ham and beef and wading to liberalism through tumblers of beer" (Mittra, 1887, p. xxxi)—allegedly "threw roast beef into a neighbouring Brahmin's house, shouting to the horrified inmates, 'Beef! Beef!'" (qtd. in Chaudhuri, 2000, p. 431). In Parker's poem, the radical Hurry Mohun tries to persuade his reluctant Brahminical friend Sam Chund that beef-eating and their shared radical principles go hand in hand: "Kings, Priests, and Laws, and Creeds, are but the tools / Which cunning knaves employ to govern fools; / Down with King's Laws and Creeds then, and in chief / With any Creed prohibiting Roast Beef" (Parker, 1831, p. 3A).[10] He then suggests that beef is not just the source of British strength but also of British technology: "Roast Beef! to what do these pale English owe, / (Or rather, yellow English) that the blow / Of British arms is strong ... / Whence are their Rail Roads, which extatic joy / Inspired in Maha Raj Ram Mohun Roy? / Whence came their wealth, their power; whence Charles Fox? / All, all originated, SAM, in Ox" (p. 3A). Hurry Mohun here evokes the second news item, James Sutherland's account published in *The Bengal Hurkaru* in September 1831, one month before "A Bengal Pastoral" appeared in the pages of *The Calcutta Government Gazette* and *The India Gazette*, of Rammohun Roy's journey by train the preceding May from Liverpool to Manchester:

[9] The poem was published "From the Calcutta Government Gazette" in *The India Gazette* on October 20, 1831, under the title "A Bengal Pastoral." Parker then republished it as "Young India. A Bengal Eclogue" (Parker, 1851, pp. 223–28). For the sake of historical context, I quote from the version in *The India Gazette*.

[10] Chaudhuri shows that Hurry Mohun is almost certainly based on Radhanath Sikdar (Chaudhuri, 2000, p. 429).

But you do not form any adequate idea of the actual velocity, until one of the trains on the opposite t[r]ack passes when the speed of both being generally accelerated for the sake of effect, they are going at the rate of 30 miles an hour!! Try to conceive; it is impossible to *describe* the effect. You are told that another train is approaching—you cast your eyes in the proper direction—you endeavour to distinguish the persons who are on the other train—you cannot even see the details of the machinery, or distinguish one vehicle from another; you hear a *whish!* like the hiss of a serpent—a dense mass intercepts your view for a second—no not even that—and the train is past! Rammohun Roy was actually unable to give utterance to a word more than "Oh! Lord! astonishing." ([Sutherland], 1831, p. 3C)

Hurry Mohun's pairing of beef with both radicalism and Rammohun Roy's ecstatic joy on a train in North West England, depicted in the newspaper account as an Asiatic's sublime and ineffable experience of steam's locomotive power, presents a very different cosmopolitan perspective than Goodeve's Brahmin parson's praise of beef on the Vaporo as simply a luxury for the stomach. As Rosinka Chaudhuri has shown, for all of Parker's satirical and at times dismissive lightheartedness, he is ultimately taking seriously, "even though ironically, the cosmopolitanism of the middle-class Hindu" (Chaudhuri, 2000, p. 439).[11] Goodeve's Protestant Brahmin, with his dietary freedoms and English interjections ("egad!"), is a kind of cosmopolitan as well, but of a different cosmos—a deracinated and westernized Indian republic replete with every modern improvement and convenience, in which there is an amicable, anglicized confusion of races and classes and the Taj Mahal has been "converted into a protestant cathedral" (Goodeve, 1835, p. 149).

The "optimistic" future imagined by Goodeve (Gibson, 2019, p. 83) accords with visions of Indian nationalism in which independence will be the natural result of diffusion, modernization, and assimilation, not, as in Kylas Chunder's "A Journal of 48 Hours of the Year 1945," the eventual but inevitable aftermath of revolt. Goodeve's tale is thus in line with the "progress narratives" that Theodore Koditschek reads in James Mill's *History* and the writings of Zachary and Thomas Babington Macaulay (Koditschek, 2011, pp. 82–87, 99–150). Charles Trevelyan best articulates the liberal assumption informing Goodeve's particular progress

---

[11] Gibson, similarly, sees Parker's humor as "more affectionate than patronizing, more amusement than ridicule" (Gibson, 2011, p. 129).

narrative, suggesting that "no effort of policy can prevent the natives from ultimately regaining their independence." There are "two ways of arriving at this point," he proposed, either "through the medium of revolution" or "through that of reform." Whereas revolution would "end in the complete alienation of mind and separation of interests between ourselves and the natives," reform would result "in a permanent alliance, founded on mutual benefit and good-will" (Trevelyan, 1838, pp. 192–93). And for Macaulay and the free-traders who endorsed colonization, it was clear that "reform" meant not just the "diffusion of European civilisation among the vast population of the East" (Macaulay, 1897, p. 141), in Macaulay's words, but the circulation of commodities and the transformation of Indians into consumers of European goods. And Goodeve wasn't alone in expressing this liberal-imperial aspiration in the sweeping language of science fiction: an article in *The Sunday Times* reprinted in *The Bengal Hurkaru* in 1828 called on the Indian government "to abandon altogether a narrow system of colonial aggrandisement ... and to build our greatness on a surer foundation, by stretching our dominion over the wants of the universe" (qtd. in Stokes, 1959, p. 43). In a similar vein, Macaulay continues, "It would be ... far better for us that the people of India were well-governed and independent of us, than ill-governed and subject to us; that they were ruled by their own kings, but wearing our broadcloth, ... than that they were performing their salaams to English collectors and English magistrates. ... To trade with civilised men is infinitely more profitable than to govern savages. That would indeed be a doting wisdom, which ... would keep a hundred millions of men from being our customers in order that they might continue to be our slaves" (Macaulay, 1897, p. 141). This is the future progress that Goodeve optimistically imagines as the outcome of liberal colonization, although he does one better than Macaulay. In the year 1980, India is both a market and well-governed, but not by its own kings: it is an independent republic in which Emmeline's aunt, Mrs. Seebchunder, is among the first female members of the newly reformed parliament (Goodeve, 1835, p. 161).

The tale is therefore full of wonder not at what technology is but at what it can do, both by and beyond supplying modern improvements: in the corner of this universe that is Goodeve's British-Indian nation, the global circulation of commodities and of scientific knowledge inaugurates and resolves the marriage plot, cementing the union of west and east by mingling the pure European extraction of the Nursings with "the blood of all the Holkars" (p. 159). The resolution, furthermore, roots out vestiges of corruption and thus purifies the Indian republic by exposing the schemes of the villain, Sir Chytun Singh. All the women in the Vaporo, we are told, were "over head and ears in love with" Charles, for, in a passage that combines English, French, and anglicized Persian by way of Hindi and suggests that European fashions are just as at home in India as are European languages, "he was dressed in the modern fashion, and doubtless those renowned Dirzies [tailors], Messrs. Mirza Beg and Chuckerbutty, had the honor of fashioning his garments; his boots were clearly of the first stamp, and the tie of his cloth *superbe*" (p. 148). And in the prison break by steam balloon, Charles, "who was very fond of chemistry," solves the problem of the platinum wire protecting the interior of the prison from above by using a "powerful solvent for platinum" that "had been lately discovered in South America," of which Charles had learned "during a short residence at Valpairaso [*sic*]" (p. 174). In the end, thanks to steam trains and balloons, boots of the first stamp, and a knowledge of chemistry acquired in Chile, east and west are united; "disgraced by the discovery of his villainy" Sir Chytun Singh "died the traitor's death"; Mrs. Seebchunder is restored to her seat in parliament; and, in Charles Holkar, India gained a new "able and eloquent statesman" (pp. 182–83).

References

Baillie, Joanna, ed. (1823). *A Collection of Poems, Chiefly Manuscript, and from Living Authors*. London: Longman, Hurst, Rees, Orme, and Brown.

Bayly, C.A. (2004). *The Birth of the Modern World, 1780–1914: Global Connections and Comparisons*. Malden: Blackwell.

Beddoe, John. (1910). *Memories of Eighty Years*. Bristol: J.W. Arrowsmith.

Bodin, Félix. (1834). *Le roman de l'avenir*. Paris: Lecointe et Pougin.

———. (2008). *The Novel of the Future*. Translated by Brian Stableford. Encino, CA: Black Coat Press.

Chatterjee, Partha. (2012). *The Black Hole of Empire: History of a Global Practice of Power*. Princeton: Princeton University Press.

Chaudhuri, Rosinka. (2000). "'Young India: A Bengal Eclogue' Or Meat-eating, Race, and Reform in a Colonial Poem." *Interventions*, 2(3), 424–41.

———. (2018). "'On the Colonization of India' (1829): Public Meetings, Debates and Later Disputes." *The Indian Economic and Social History Review*, 55(4), 463–89.

De, Barun. (1975). "A Biographical Perspective on the Political and Economic Ideas of Rammhohun Roy." In V.C. Joshi, ed., *Rammohan Roy and the Process of Modernization in India.* Delhi: Vikas, pp. 136–48.

Dutt, Kylas Chunder. (1835). "A Journal of 48 Hours of the Year 1945." In D.L. Richardson, ed., *The Calcutta Literary Gazette or Journal of Belles Lettres, Science, and the Arts,* new series 3(75) (June 6), 355–56.

"A Family Conversation." (1834). In D.L. Richardson, ed., *The Calcutta Literary Gazette or Journal of Belles Lettres, Science, and the Arts,* new series 2(49) (December 6), 353–56.

Gibson, Mary Ellis. (2011). *Indian Angles: English Verse in Colonial India from Jones to Tagore.* Athens, GA: Ohio University Press.

———, ed. (2019). *Science Fiction in Colonial India, 1835–1905: Five Tales of Speculation, Resistance and Rebellion.* London: Anthem Press.

Goodeve, Henry Hurry. (1835). "1980." In David Lester Richardson, ed., *The Bengal Annual. A Literary Keepsake for MDCCCXXXV.* Calcutta: Samuel Smith, pp. 147–83.

———. (1836). "The Doctor's Tale." In David Lester Richardson, ed., *The Bengal Annual. A Literary Keepsake for MDCCCXXXVI.* Calcutta: Samuel Smith, pp. 58–93.

———. (1847). "Hindoo Medical Students." *The Lancet,* 49(1224), 189–90.

Grant, Colesworthey. (1837). "H.H. Goodeve Esqre. M.D." *The India Journal of Medical and Physical Science,* new series 2(7) (July 1), facing 429.

———. ([1850?]). *Lithographic Sketches of the Public Characters of Calcutta. 1838–1850.* Calcutta: W. Thacker and Co.

Hazlitt, William. (1890). "The New School of Reform; A Dialogue Between a Rationalist and a Sentimentalist." In William Carew Hazlitt, ed., *The Plain Speaker: Opinions on Books, Men, and Things.* London: George Bell and Sons.

Koditschek, Theodore. (2011). *Liberalism, Imperialism, and the Historical Imagination: Nineteenth-Century Visions of a Greater Britain.* Cambridge: Cambridge University Press.

Macaulay, Thomas B. (1897). "Government of India." In Hannah More (Macaulay), Lady Trevelyan, ed., *The Works of Lord Macaulay: Complete.* Vol. VIII. New York: Longmans, Green, and Co, pp. 111–42.

Mittra, Kissory Chand. (1870). *Memoir of Dwarkanath Tagore.* Calcutta: Thacker, Spink and Co.

———. (1887). Appendix B, "The Hindu College and Its Founder," in Peary Chand Mittra, *A Biographical Sketch of David Hare.* Calcutta: W. Newman, pp. x–xxxxi.

Parker, Henry Meredith. (1831). "A Bengal Pastoral." *India Gazette,* October 20, 3A.

———. (1851). *Bole Ponjis. Containing The Tale of the Buccaneer; A Bottle of Red Ink; The Decline and Fall of Ghosts; and Other Ingredients*. Vol. 2. London: W. Thacker.

Parkes, Fanny. (2001). *Wanderings of a Pilgrim in Search of the Picturesque*. Edited by Indira Ghose and Sara Mills. Manchester: Manchester University Press.

Sarkar, Abhishek. (2023). "The Celebration of Technoscience in '1980,' a Futuristic Science Fiction Published from Calcutta in 1835." *ANQ: A Quarterly Journal of Short Articles, Notes and Reviews*, 36(1), 45–50.

Sarkar, Sumit. (1975). "Rammohun Roy and the Break with the Past." In V.C. Joshi, ed., *Rammohan Roy and the Process of Modernization in India*. Delhi: Vikas, pp. 46–68.

Sartori, Andrew. (2008). *Bengal in Global Concept History: Culturalism in the Age of Capital*. Chicago: University of Chicago Press.

Sen, Asok. (1975). "The Bengal Economy and Rammohan Roy." In V.C. Joshi, ed., *Rammohan Roy and the Process of Modernization in India*. Delhi: Vikas, pp. 103–35.

Stokes, Eric. (1959). *The English Utilitarians and India*. Oxford: Clarendon Press.

[Sutherland, James.] (1831). "London, 6th May, 1831." *Bengal Hurkaru*, September 10, 3C-D.

Trevelyan, Charles E. (1838). *On the Education of the People of India*. London: Longman, Orme, Brown, Green, & Longmans.

White, Daniel E. (2013). *From Little London to Little Bengal: Religion, Print, and Modernity, 1793–1835*. Baltimore: Johns Hopkins University Press.

———. (2016). "'Zig Zag Sublimity': John Grant, the Tank School of Poetry, and the *India Gazette* (1822–1829)." In Rosinka Chaudhuri, ed., *A History of Indian Poetry in English*. Cambridge: Cambridge University Press, pp. 147–61.

Whitfield, Michael. (2010). *Dr Goodeve and Cook's Folly*. Bristol: Avon Local History & Archaeology.

———. (2022). "Henry Hurry Goodeve (1807–1884), the First Professor of Anatomy in India." *Journal of Medical Biography*, 30(1), 6–14.

Wordsworth, William. (2004). *Sonnet Series and Itinerary Poems, 1820–45*. Edited by Geoffrey Jackson. Ithaca: Cornell University Press.

Zastoupil, Lynn. (2010). *Rammohun Roy and the Making of Victorian Britain*. New York: Palgrave.

CHAPTER 5

Henry Meredith Parker and the Miserable Hour of a World's Desolation

Abstract A different liberal-imperialist vision of technology, global capitalism, and British-Indian nationalism emerges from the activities, friendships, and writings of Henry Meredith Parker, a prominent figure in Calcutta's commercial, literary, theatrical, and social scenes. Presenting original research fleshing out his life and career, I trace his friendship with Dwarkanath Tagore, who in 1833 became a Director of the New Bengal Steam Fund, of which Parker was Chairman. Relying on a hitherto undiscussed source in which Parker describes himself as "differing entirely from Mr. [James Silk] Buckingham in politics," the chapter presents Parker as a nonpartisan patriot and a loyal Company servant. Parker's politics and sociability merge in "Calcutta Pococurante Society" (1833), the curious proceedings of a mystic dinner club. From the Society's debate over "Utilitarianism *versus* poetry," I turn to Parker's "tale of the nothingness of man," in which the world as we (will) know it ends on May 10, 2054. Parker's and Goodeve's visions of a future liberated by machines differ in more than just the opposite outcomes of technological dominion they depict. Parker's tale, I argue, is a Burkean response to reckless transformations of the earth in the service of maximal postnationalist global capitalism.

Keywords Henry Meredith Parker · "The Junction of the Oceans" · Dwarkanath Tagore · James Silk Buckingham · "Calcutta Pococurante Society" · Edmund Burke

D. E. White, *Romanticism, Liberal Imperialism, and Technology in Early British India*, https://doi.org/10.1007/978-3-031-60705-9_5

A different liberal-imperialist vision of technology, global capitalism, and British-Indian nationalism emerges from the activities, friendships, and writings of Parker, a "man of rare talents and brilliant attainments" (Stocqueler, 1873, p. 89) and a prominent figure in the commercial, literary, theatrical, and social scenes in Calcutta during the 1820s and 1830s. Looking back from the perspective of 1859 in the face of hostile questioning from a Select Committee of the House of Commons seeking to pin blame for the Rebellion on Company misrule and to give all credit for progress to "persons unconnected with the Company," Ross Donnelly Mangles, Chairman of the East India Company in 1857 and then member of the new Council of India after its establishment in 1858, was pressed to name supporters of steam communication among Company servants, as opposed to nonofficial "settlers, commercial people, and English capitalists": he replied, "Mr. Henry Meredith Parker, a high functionary of the Government, and many others I could name [he didn't], aided the enterprise to the utmost of their power" (*Reports*, 1859, pp. 131–32).

Parker (Fig. 5.1) was born on June 4, 1795, into a remarkable performance family as the son of the equestrian and pantomime performer and producer William Parker and his second wife, the "celebrated *danseuse*" (Stocqueler, 1873, p. 89) Sophia Granier, scion of an extended clan of Covent Garden dancers. Parker's half-sister (daughter of William Parker and his first wife) was the famous actress Nannette Parker, who was married to the still more famous actor Henry Erskine Johnston from 1796 until their separation around 1811 (Highfill, 1982, pp. 201–202). Parker was educated, according to his Petition to the Court of Directors in early 1814, by "the Rev. Dr. Williams of Soho Square Academy" ("The Humble Petition," 1814, p. 19), where "drama was seen as central to education, with at least one major production of a tragedy or comedy being mounted each year" (Cox, 1999, p. 307), and where numerous theatrical families enrolled their sons. He seems to have left the academy at age 13, "when a mere child, he became a servant … in H.M. office of ordnance," finding himself by 1812 in the Army of Sicily as an officer in the Commissariat during the Peninsular War: "He was thrice at Algiers, and at one time for many months domiciled with a Greek pirate, a vice consul at the wildest and most cut-throat place on the whole coast of Barbary" ("H.M. Parker," 1840, p. 589). He was "Said in his youth to have been a violinist at Covent Garden Theatre" (Buckland, 1906, p. 328; cf. Corfield, 1913, p. 436), either at a very early age indeed, before he was 13, or in the brief period between his return from the

Fig. 5.1 "H.M. Parker, Esq.," by Grant, from *Lithographic Sketches of the Public Characters of Calcutta* ([1850?]), Plate 58; from the British Library Archive shelfmark V 6602[1]

[1] This lithographic sketch was originally published in "H.M. Parker" (1840, facing p. 589).

Continent and his departure for India in early 1814. His family clearly had powerful connections: Henry Erskine Johnston's godfather was Thomas Erskine, Lord High Chancellor in 1806–1807 (Campbell, 1996, p. 22), and Stocqueler suggests that the appointment to the Commissariat was procured "by some powerful friend (Lord Moira, I think)" (Stocqueler, 1873, p. 89). It was in all likelihood through the influence of Lord Moira (Francis Rawdon-Hastings, Marquess of Hastings), Governor-General of Bengal from 1813 to 1823, that Parker's Petition to the Court of Directors for the position of Writer on the Bengal Establishment was presented to the Court in December 1813 by none other than "H.R.H. The Prince Regent" ("The Humble Petition," 1814, p. 14).[2] Having arrived in India on July 30, 1814, Parker rose through the ranks of the Salt Department, eventually (following a return to Europe on medical leave from 1824 to 1827) serving as Secretary to the Board of Customs, Salt, and Opium in 1828 and ascending to First Member of the Board in 1838.

Parker, Plowden (whom we encountered before among the members of the Agricultural and Horticultural Society), and Dwarkanath were close friends and business associates. While Plowden was Superintendent of the Western Salt Chowkies (custom houses established as part of a system to prevent smuggling) in July 1816, Parker became his Assistant, succeeding him as Superintendent that November when Plowden became Officiating Secretary to the Board of Trade (Doss, 1844, pp. 277–78). Both continued to serve in various capacities connected to the salt trade, but their association was more intimate. With Parker present, Plowden and his wife Frances stood as witnesses at the weddings of both of Parker's sisters, Sophia Zenana and Josephine; whereas Josephine's

---

[2] On Lord Moira's patronage, see Buckland (1906, p. 328) and Corfield (1913, p. 436). Margaret Makepeace has suggested the possibility that the Prince Regent may have assisted Parker as a way of appeasing Nannette Parker and Henry Erskine Johnston and limiting the fallout of scandal (Makepeace, 2021, n.p.). During a performance at Drury Lane, the Prince Regent seems to have forced his way into Nannette Parker's dressing room and sexually assaulted her, "press[ing] his gallantries so far" that her husband, who followed the Prince into the room, "horsewhipped him there and then" (Baynham, 1892, p. 87), an assault for which he was never prosecuted. Baynham dates the incident to 1814, but by then the couple had been separated for several years (their divorce was formalized in 1820). (Campbell suggests that the incident took place when Lord Erskine was Lord Chancellor [1806–1807] and that it was through Erskine's influence that Johnston avoided prosecution [Campbell, 1996, p. 22].) In light of Lord Moira's earlier patronage, I find it more likely that it was Moira's influence with the Prince Regent that led to the Regent's support for Parker's appointment.

marriage (in 1832) took place in Calcutta's St. John's Cathedral ("Marriages," 1832, n.p.), Sophia Zenana's marriage (in 1822) took place at the Plowden's home in Noakhali, in present-day Bangladesh ("Dacca Marriages," 1822, n.p.), where Parker was then serving as joint magistrate (Boase, 1965, p. 355). After Plowden died in 1836, Parker and Frances married on Christmas Day 1837, with Dwarkanath among the six witnesses. Dwarkanath became the major patron of the Chowringhee Theater (Kling, 1976, p. 49), of which Parker was the director, purchasing it in 1835. Kling describes the Theater as "a setting for social intercourse between the races" (p. 160) with, according to Emma Roberts, "parties of Hindostannee gentlemen beautifully clad in white muslin ... sitting as near the stage as possible" (qtd. p. 160). When Emily Eden joined her brother, the new Governor-General of India George Eden, 1st Earl of Auckland, on a visit to Dwarkanath's villa in August 1836, she recorded that "Dwarkanauth talks excellent English, and had got Mr. Parker, one of the cleverest people here, to do the honours" (Eden, 1872, p. 216). Parker and Plowden frequently wrote to Dwarkanath using the language of intimacy, a subject to which we will return. Before his departure in 1836 (he would die at sea), Plowden wrote, "My Dear Dwarkanath,— Kindest friend, there is no one I leave with more regret than I do you ... Mr. Parker leaves me to-night. God bless you, and let me hear from you soon" (Mittra, 1870, p. 22), while Parker acknowledged Dwarkanath's donation of a lac of rupees (approximately £6,500) to the District Charitable Society—the funds were earmarked for the relief of "the poor blind," and Parker was one of the fund's trustees—in 1838 thus: "I have known you now, my dear Dwarkanath, for many years. I have known and revered your ... unequalled kindness of heart ... Need I say more to justify my anxious desire ever to be considered your affectionate and sincere friend" (Mittra, 1870, p. 22). Upon Parker's final return to England via Suez in 1842, Dwarkanath accompanied him on the *India*, a steamship owned by Dwarkanath.

Both Plowden and Parker were supporters of Dwarkanath in the 1820s, backing his promotion to the post of dewan of the Board of Customs, Salt, and Opium, "one of the three or four most important positions an Indian could hold in the government service" (Kling, 1976, p. 37). After Dwarkanath resigned from the Board and founded Carr, Tagore, and Company in 1834, becoming the first Bengali "merchant" as a partner in a trading company, as opposed to a "banian" in the employ

of Europeans (Kling, 1976, p. 76), he saw an opportunity in the deregula-
tions of the new Charter Act as well as the fallout of the financial crisis. At
the end of 1832, the crisis took down Alexander and Company, the firm
contracted by the East India Company to manage Raniganj, a massive coal
field that fueled the engines of Bengal and has yet to be exhausted today.
In 1836, in "the most important single transaction of Dwarkanath's business career," Carr, Tagore, and Company purchased the contract from
Alexander and Company, giving Dwarkanath "virtual control over the
supply of fuel in the Bengal presidency" (Kling, 1976, pp. 94–95). In
order for Raniganj to become "the heart of Dwarkanath Tagore's business empire" (Kling, 1976, p. 94), however, he needed further contracts
to supply the coal he now controlled to the government, especially to
government vessels, and for those contracts he required the approval of
the Marine Board, of which Parker had been a member since 1833, the
same year in which Dwarkanath and Parker joined the New Bengal Steam
Fund.

Beyond his intimacy with Dwarkanath, Parker sustained an unusu-
ally wide range of friendships across lines of race and gender. The
Chowringhee Theatre, as we have seen, was a site of diverse social intercourse, and "No dinner party of Parker's," it was said, "was complete
without Derozio," the "East Indian" (i.e., mixed-race) poet, teacher, and
journalist "who added a gift for brilliant conversation to his youthful
eagerness and charm" (Dover, 1937, pp. 110–11). Mary Ellis Gibson
and Rosinka Chaudhuri suggest that "Parker's dinner table was a meeting
place for British writers and their Indian interlocutors" (Gibson, 2019,
p. 32). Derozio wrote a "Sonnet to H.M. Parker, Esq.," and the poet and
travel writer Emma Roberts was a mutual friend of the two, while Hindu
College student Kasiprasad Ghosh dedicated his poem "The Shaïr" to
Parker.[3]

Parker's politics were slippery. In fact, nonpartisan loyalty to the East
India Company seems to have been among his central tenets, regardless of
his embrace or rejection of policies associated with antagonistic camps. An
excellent starting point emerges from the list of signatories appended in
1826 (during Parker's medical leave) by Buckingham to his petition to the
House of Commons for redress in consequence of losses incurred upon
his expulsion from India three years earlier. Alongside the likes of "Jeremy

---

[3] On Kasiprasad, see Chaudhuri (2002, pp. 60–80) and Gibson (2011, pp. 137–54).

Bentham, esq. Bencher of Lincoln's Inn," "Sir Francis Burdett, bart. M.P. Westminster," and "Henry Brougham, esq. M.P. Winchelsea," Parker's is the only signature to bear more than a name, rank, occupation, and residence, reading, "Henry Meredith Parker, esq. late of Bengal—who adds, after his signature, the following sentence:—'Differing entirely from Mr. Buckingham in politics, but convinced that he is a sufferer for conscience sake; and, by an intercourse of ten years, in India and in England, that he is an upright, honourable, and excellent man'" ("Mr. Buckingham's Petition," 1826, n.p.).

Buckingham was first and foremost a free-trader opposed to the East India Company's (or any) commercial monopoly, and he supported a free press in India, judicial reform (to introduce juries and diminish the power of individual judges), Rammohun's agendas for native education and newspapers, the abolition of slavery and sati, and various measures to curtail what Buckingham saw as the despotic power of the Company, especially by granting Britons access to India without requiring a license and giving Britons the right to own land and trade privately in India.[4] In many ways, then, he and Parker would have seen eye to eye, and not just on the freedom of the press. Loyal Company servant that he was, Parker's chairmanship of the New Bengal Steam Fund nonetheless placed him in line with the Calcutta merchant lobby and opposed to "metropolitan control over local enterprises" (Kling, 1976, p. 58). Both Buckingham and Parker were on good terms with likeminded members of the *bhadralok* (Rammohun and Dwarkanath), while Buckingham's *The Calcutta Journal* and (back in London) *The Oriental Herald* regularly published Parker's poetry. Both were appalled by evangelicalism, with the latter ironically dedicating his later *Caste and Conversion* (1858) "To that devout, earnest, and conscientious body of Englishmen, whose

---

[4] Chaudhuri points out that Buckingham and the free-traders rhetorically linked "despotism" to monopoly, thus opposing despotism as an infringement upon the rights of British merchants and (especially) indigo planters, not upon the freedoms of Indians themselves (Chaudhuri, 2018, p. 470). An anonymous mixed-race Indian writer ("S.J.") in Derozio's magazine *The Kaleidoscope* for September 1829 accordingly opposed colonization as a land grab and instead supported "admitting natives and Indo-Britons to a participation of privileges ... with the Europeans" ("Indo-Britons" means mixed-race Christians, also called "East Indians," such as Derozio and the anonymous writer), going so far as to raise the specter of "discontent ... brooding into rebellion" even as he acknowledged that colonization would foster "the introduction of the arts and sciences of Europe" and advocated for institutions similar to the Hindu College to be established "at every town in the *mofussil*" ("On the Colonization," 1978, pp. 33–34).

fervent zeal for Conversion has clearly helped to create a fearful Mutiny, and will probably excite a National Rebellion in India, thereby throwing back the sacred cause of Christianity in that country for centuries" (Parker, 1858a, p. 4). Parker's *The Empire of the Middle Classes* (1858), in which the titular phrase is juxtaposed to the "Empire of the House of Commons" (Parker, 1858b, *passim*), evinces his longstanding hatred of patronage, jobbery, and nepotism and his fears that under Parliamentary control the government of India will be subject to the whims of the latest party in power, which is to say to the latest political system: "One day the 'parti prêtre,' which appears to have a personal quarrel with Vishnoo, will be in power, and drive our unhappy dusky fellow-subjects to despair by pushing proselytism to the verge of persecution"; then the "'parti philosophe'" will reverse the trend and proclaim absolute liberty of conscience; next, another set, "haughty, imperial, all for absorption, annexation, conquest, if necessary, and complete subjugation" will succeed; and finally, a party breathing "nothing but peace and love" will "utterly repudiate the stern views of their predecessors, and hasten, perhaps, to restore pensioned sultans, and unemployed rajahs in 'a transport of republican enthusiasm and universal philanthropy,' never equalled since the days of the needy knife-grinder" (Parker, 1858b, pp. 13–14). Above all, no one on this merry-go-round of patronage and systematic politics will have any experience of India. His *A Plan for the Home Government of India. With Provisions Calculated to Prevent or Limit the Dangers and Evils of Patronage* (1858) proposed to replace the Parliamentary Board of Control and Company Directors at Leadenhall Street. This plan similarly advocated for local knowledge and experience in the government of India, which would be led by a "Great Council of India" comprising twenty-one elected counselors, fifteen European and six Indian, with a residency requirement for European elected members of eighteen years in India after the age of twenty and for European electors of fifteen years (Parker, 1858c, pp. 7–10). Buckingham would have broadly agreed with Parker's support for home rule but would have called for an autonomous local government over continued Company control. (He also would have been quick to point out that corruption had always played a far greater role in Company advancement than was allowed by Parker, who after all owed his own appointment to unusual patronage.)[5]

---

[5] On Company patronage, see Muir (2019, pp. 283–310).

Parker did differ entirely from Buckingham in another respect, however, and his signature represents an attempt to distance himself from both his co-signatories and Buckingham the "radical reformer" associated with the likes of William Hone, Richard Carlile, and Henry Hunt and a sworn enemy to all sorts of Old Toryism (embodied in India by John Adam) and John Bullism. Parker saw himself, on the other hand, as above all a nonpartisan patriot and a loyal Company servant. When, almost a century before Gandhi's march, the Government's enduring salt monopoly came under attack in 1836 due to its exploitation of the molunghees (malangīs, coastal salt manufacturers), whose labor critics of the Company described as "coerced,"[6] Parker defended the monopoly, standing forth, in the words of an 1840 sketch of his life, as "the Government's shield" and wielding "the Government's sword of defence" ("H.M. Parker," 1840, p. 589). Even as both Buckingham and Parker were essentially liberal imperialists in favor of free trade (flexibly so in Parker's case), the freedom of the press, home rule, and an empire maintained by commerce, communication, and opinion rather than by force, Parker could and did make the kind of toast that would have turned to ash in the mouth of Buckingham. At the third Free Press Dinner at the Town Hall of Calcutta, in February 1838, Parker stood and proposed:

> Gentlemen; my toast is the British Army. (Cheers.) I know there has been discussion infinite touching the politics of the British Army. Whether it was Whiggish or Toryish, Reformatory or Conservative—whether it loved a Free Press or did not love a Free Press—for my own part, I will own to you candidly that I don't care a fig what its politics are, or what its feelings are, on the question I have hinted,—it is sufficient for me to know, that through long years of peril and gloom the British Army fought and bled, that the hearths and altars of their country might not be polluted by a foreign foe. (Cheers.) ("Free Press," 1838, pp. 83–84)

Parker then recommended a celebration of Metcalfe separate from the commemoration of the 1835 Press Act in the form of another dinner "distinct from all political feeling ... to shew their regard for a great and good man, whose heart was open as day to melting charity, and whose hand was as open as his heart" ("Metcalfe Meeting," 1838, p. 94).

---

[6] See Serajuddin (1978) and Ray (2011, pp. 133–70).

This language of nonpartisan friendship, in fact, was part and parcel of Parker's politics of sentiment. A further and fascinating source on Parker's sociability appears in the form of a curious text appended to the beginning of the October 1833 issue of David Lester Richardson's short-lived *The Calcutta Quarterly Magazine and Review* (Chanda, 1987, p. 159). "Calcutta Pococurante Society," provides the rules and proceedings of a mystic dinner and drinking club, the members of which went by occult titles and nicknames.[7] The society's insignia was "an alligator *rampant* trampling on the new moon" (Fig. 5.2), the "mysterious meaning" of which was to be "communicated to Ministers and blue Teraphs only" ("Calcutta," 1833, p. 4). In a copy bearing the stamp of the Uttarpara Jayakrishna Public Library in West Bengal, a manuscript hand has identified five of the six members of the "Secret Committee for the first golden circle," with the "Archialligatros," Zegri de Rohan, identified as "HMP" for Henry Meredith Parker. Other members include the "Mystics" Wilfred, who is identified as "DLR" for David Lester Richardson; Lincoln, "WCH" for W.C. Hurry (member of the Agricultural and Horticultural Society, and Goodeve's first cousin); Candide, "Theo Dickens" for Barrister of the Supreme Court Theodore Dickens (member along with Rammohun and Dwarkanath of the Calcutta Unitarian Committee in the late 1820s); and Glengyle, "Short," whom I have been unable to identify (p. 4). (Parker, Hurry, and Dickens all wrote for Richardson's *The Bengal Annual*.) The Secret Committee meets at night in "A Turkish tent of white silk," inside of which "there is a cool and fragrant air … produced by Altenberg's refrigerator," and the viands include "two iced Melons …

---

[7] See Ehrlich (2016). Like Ehrlich, for whose correspondence I am grateful, I have only been able to consult a PDF of the copy stamped with the mark of the Uttarpara Jayakrishna Public Library. The pagination of *The Calcutta Quarterly Magazine and Review* for October 1833 is curious: under "Original Articles," the Contents first list "Proceedings of the Calcutta Poco-curante Society" with the page numbers provided in brackets, thus: "[1–60]" (the text itself is also paginated in brackets, which I omit in my citations). After a line divider, the pagination then recommences at p. 1, without brackets ("Prefatory Remarks, 1"; "The Song of the Forge, by H.M.P., 2," etc.). It appears, therefore, that the issue of *The Calcutta Quarterly Magazine and Review* was printed first and separately from "Calcutta Pococurante Society," which was then printed (with its own bracketed page numbers) and appended. Without being able to examine the physical copy, I am unable to tell if the two were bound together or if "Calcutta Pococurante Society," was subsequently sewed or (less likely due to length) tipped in.

Fig. 5.2 Alligator *rampant* trampling on the new moon, from "Calcutta Pococurante Society" (1833)

beaming rich and ripe through a glittering incrustation of frosty parti-cles" (p. 5).[8] The "Rules" set forth the subjects to be discussed: after "Three courses and a desert" followed by "An additional course," the society will turn to meatier matters: "Cant, Humbug, and Absurdity in all their branches whether Tory, Whig, Radical, Ultra, or Liberal ... From the absurdities of Utilitarianism to the absurdities of Romance" (p. 3). Parker's humor and nonpartisanship permeate the piece, which I strongly suspect he wrote: after the appended text, the magazine opens with "Prefatory Remarks" followed immediately by Parker's poem "The Song

[8] I am unable to identify Altenberg, but the refrigerator is almost certainly a therman-tidote: the air in the tent is not just cool but also "fragrant," and Parkes, describing the thermantidote, emphasizes "how fragrant ... is the scent of the fresh khās khās" (Parkes, 2001, p. 149). See above, Chapter 4, note 3.

of the Forge."[9] Notably absent from the kinds of humbug and absurdity to be subjected to the society's scrutiny and sarcasm are religious and Company cant. And the reason for refraining "from any comments on Religious cant" is that, "while throwing stones at those meretricious gauds … with which human folly has disfigured the simple and beautiful edifice of Christianity," the society's "attacks might be misconstrued into assaults upon the structure itself." Regarding Company cant, as ever for Parker loyalty remained paramount: "if it were possible for the Government of British India to use cant or employ humbug, this Society would respect it for obvious reasons" (p. 3).

After a long discussion of Shelley and techniques for cooling claret and hock, with detailed menus (in French) and wine lists interspersed among the conversations, the second meeting of the Secret Committee (in which the original six are joined by four others) considers "that famous controversy … which blazed up for a time in this good town of Calcutta with so much fierceness and energy … I mean Utilitarianism *versus* poetry" (p. 38). De Rohan/Parker asks, "what was it all about? *Wilfrid*, my good fellow, pray what *do* the Utilitarians say touching the merits of poetry. I don't understand either Coptic, Chinese or Benthamic, so kindly edify us by giving the opinion of the great sage and his followers on the point at issue" (p. 38). Wilfrid quotes the notorious passage in which Bentham compares poetry with the game of pushpin, arguing that if pushpin furnishes more pleasure, it is of greater utility than poetry, and that because everyone can play pushpin, whereas poetry is relished by only a few, pushpin is the more valuable of the two entertainments. De Rohan, who counts himself among the "idealogist-despising 'men of this world'" (p. 43) and is ever invested in concrete experience over abstraction, characteristically interjects, "I'll bet a sixpence, there's not an Utilitarian in the room, who knows what Pushpin is, or can play at it. I am accomplished both in theory and practice" (p. 39). A long

---

[9] In the discussion of refrigeration techniques, Glengyle embeds references to Parker's *The Draught of Immortality* (1827): "The old abdar [domestic servant who cooled beverages] fashion was a regular churning process, and when the Amreeta was handed up, it had about as much gusto as the churned ocean, but alas! there was any thing but immortality in the draught" ("Calcutta," 1833, p. 34). And it could be either a coincidence or an inside joke, but "the Southern Cross," which is the "sign" of "the first golden circle" in "Calcutta Pococurante Society," (pp. 4, 24, 57), also happens to be the final sign in "The Junction of the Oceans," the "perfect cross" in the south-east that suddenly appeared on "the day of the world's calamity" (Parker, 1835, pp. 54–55).

debate ensues regarding whether Bentham was comparing the form or the spirit of poetry to the game of pushpin, and to settle that debate, the society realizes it needs a working definition of poetry, fearful as that challenge is. Another member (Le Voyageur) having proposed that poetry "represents or describes human passions, and ... is capable of exciting in us, emotions similar to those produced by the passions themselves," De Rohan cites the Marseillaise and Ca Ira (p. 42) as songs that moved multitudes, surpassing even pushpin in the production of passion. Candide then "*Whistles Lillibullero in a low sweet whistle,*" presumably, like dear Uncle Toby, to indicate the absurdity of the whole conversation, but De Rohan reads his interruption completely otherwise. Responding, "Yes, I understand you," he hails the "gabble of geese which drove James from his three kingdoms" as "positively sublime" and seconds Candide's whistled implication that Lillibullero should be added to the list of revolutionary songs that prove "the utility of poetry" (p. 42). He then pronounces that, on the basis of this judgment, he should be understood "as a *practical* Utilitarian, not as a Benthamite" (p. 46). Recall that for Bodin's Anti-Prosaic Association coal was a powerful agent of a sad, uniform, and monotonous civilization destructive of all poetic life. A far cry from either sad and prosaic uniformity or the air-conditioned Vaporo in "1980," on which "the republican spirit of the time allowed of no distinction," the refrigerated Turkish tent of the "Calcutta Pococurante Society," houses an anti-utilitarian sociability in which alcohol, food, poetry, and wit nourish an idealogist-despising, little-caring, but cosmopolitan erudition that values practical experience and knowledge over abstract theory and cant, whether "Tory, Whig, Radical, Ultra, or Liberal."

Although Joshua Ehrlich is right to signal the work's "clubby parochialism" and its relegation of the "everyday experience of life in India ... to outside the tent flaps" (Ehrlich, 2016, n.p.), the fact of the jeu d'esprit's restriction of a drinking party to Europeans should not obscure the Archialligatros' commitment here to everyday experience (of pushpin, of Calcutta controversies) and, in his other writings, to what Parker calls the "East Easty" (Parker, 1851b, p. 139). The phrase comes from his collected works of 1851, in which the introduction to the section titled "Orientalisms" opens, "I should be a very alligator of ingratitude towards the benevolent occidental, or accidental, reader who may, through great good nature, have accompanied me thus far ... if I did not warn that kind hearted individual of what is to come." And what is to come will not be an Orient replete with "Gardens of gul in her bloom" and lands "Where all but the spirit of man is divine," but rather the "simple prosaic East of this everyday world," or "no such East as the reader has, probably, been familiar with" (Parker, 1851b, p. 139). Rosinka Chaudhuri describes the satiric and comedic elements of Parker's poetry as antecedents to Kipling's Indian verse, both characterized by "an attention to detail and the dailiness of colonial life" (Chaudhuri, 2002, p. 187).[10] A prolific writer, Parker published poetry and prose under his own name or initials as well as a range of pseudonyms, including "Bernard Wycliffe" and "Dionysius Stubbs." Among his best-known works was "The Draught of Immortality" (1827), based on a theme from the *Mahabharata* and precisely the kind of poem that was falling out of fashion with those who, in the wake of Mill's *History*, sought to disparage Indian history and culture: "in homage to William Jones' long poem 'The Enchanted Fruit,'" Chaudhuri writes, "and in the manner of Southey and Moore, it incorporated many Hindu mythological motifs and much Sanskrit etymology elucidated in footnotes" (Chaudhuri, 2002, pp. 36–37).

On May 10, 2054, "that fatal, that tremendous day" (Parker, 1835, p. 2), the world as we (will) know it will end—so imagines "The Junction of the Oceans," "a tale of the nothingness of man" (p. 5) published in Calcutta in 1835 and told in Panama in 2074 by one of the few survivors of the cataclysm, Carlos, to his son Alfred.[11] The plot is simple. Carlos relates the events leading up to and including the "unparalleled disaster" (p. 3) that took place twenty years earlier, when, from his hunting lodge high in the mountains above the city of New Panama, he witnessed "gigantic machinery" (p. 34) complete the final cut and open a vast canal, wide enough for "ten thousand vessels, from the gigantic steam-ship to the humble coaster and the humbler fishing boat ... to pass through" (p. 32). Surviving along with him are Domingo, "a fine young negro" (p. 26), and Domingo's mother Sarah. Carlos then describes the ensuing tsunami, which took the life of his wife. Thereafter, he discovered a giant ship washed up on what is now the shore below his lodge. On board was a party of three survivors, including young Henrietta Albany, an Englishwoman who will become Alfred's mother, and two "natives of the

---

[10] See as well White (2013, pp. 142–44) and (2016), and Ní Fhlathúin (2015, pp. 28–30).

[11] In the revised version published in *Bole Ponjis* (1851), Parker pushes the dates of the apocalypse and the telling of the tale back to 2076 and 2098, respectively (Parker, 1851a, pp. 132, 135).

Phillipines ... dressed in the picturesque costume of the Manilla Creoles" (p. 49), one old and one young, the latter of whom will become the mother of Domingo's children. Carlos finally tells Alfred that signs in the heavens and his own heart are calling him to fulfill "a high commission to seek if there yet survive in islands like our own some remnants of the human race" (p. 51). The tale ends as he prepares to depart, along with Domingo, three of Domingo's sons, and two of Alfred's brothers, on this mission to resume human civilization.

While planning the canal, like so many Victor Frankensteins the engineers in their presumption had dismissively asked, "what is there to apprehend? The respective levels of the two oceans have been measured to the fraction of a line, and to fancy that any material geological ... change can result from their union is a demonstrable absurdity" (p. 16). This apprehension was not the product of Parker's imagination. In 1824, Buckingham's *The Oriental Herald* published Parker's "Leaves and Flowers, or the Lover's Wreath" immediately before an article "On the Proposed Communications between the Atlantic and Pacific Oceans." So if Parker had looked for his poem in print (not an unusual thing for an author to do!), on the very next page he could have seen that title and gone on to read:

> A question of serious importance has received considerable elucidation in the course of the discussions on this topic; namely, whether of two neighbouring seas, as is the vulgar opinion, the one is more elevated than the other. This opinion has been so prevalent, that some have even gone so far as to predict that the consequence of opening a communication between the two seas, would be the inundation of the entire isthmus. ("On the Proposed," 1824, p. 411)[12]

Not just of the entire isthmus but of the entire world, the ensuing inundation in "The Junction of the Oceans" marks "the desolation of mighty empires" (Parker, 1835, p. 3), inflicting "a terrible chastisement on the proud presumption of man" (p. 2).

---

[12] Cf. Erasmus Darwin's conjecture that if the Straits of Gibraltar were "opened by an earthquake in antient times," then the myth of the flood of Deucalion could have originated in the "immense current of water" that would have "run into the Mediterranean from the Atlantic" due to the difference between the surface levels of the two (Darwin, 1791, pp. 30–31).

The project was going to be the crowning utopian, and utilitarian, achievement of liberal imperialism. A century of warfare having been closed by the Peace of Frankfort, "new European republics, every where triumphant over the misrule and oppression of the past, formed a consolidated and harmonious system" (p. 5). Peace unleashed "schemes for ameliorating the condition of society," which, "especially through the medium of that chiefest ameliorators commerce, were in rapid circulation, while the general spread of science and information enabled men" to pursue plans "for the general welfare" (pp. 5–6). The question became "how to turn to the greatest advantage of mankind, that vast power of intelligence, and that prodigious energy which the general pacification had released from the stern bondage of war and politics" (p. 5). And the answer was maximal postnationalist global capitalism and the radically anthropocenic transformation of the earth to facilitate it: Great Britain and France "were united by an artificial isthmus covered with flourishing towns," while "A similar connection spanned the Hellespont, and there brought Europe into contact with Asia" (p. 6). The Isthmus of Suez was cut through: "the coral rocks of the Red Sea were scattered by explosive forces, or removed by mechanical powers, unknown to the philosophers, even of the twentieth century. Majestic fleets stemmed alike the winds and currents, and within a lunar month and a half the produce of Norway or of Holland was landed on the marble quays of the cities of the Indus" (p. 6). One great project remained: the "union of the Pacific and Atlantic Oceans" (p. 7), and so "The wealth of Europe, of Asia, and the Americas was lavishly poured forth to accomplish this crowning work of civilization" (p. 11). The Panama Canal would be "the last connecting link which will so closely unite mankind that the most opposite kingdoms of the globe shall be less remote from each other than were the neighbouring states of the last ages. Here is that which shall communicate civilization and science, the products and the intelligence of nations" (p. 12). The canal will be "a new highway for the march of civilization and human happiness," an achievement of "boundless utility" (p. 22).

Parker's and Goodeve's visions of a future liberated by machines differ in more than just the opposite outcomes of technological dominion that they depict. Whereas the Vaporo in "1980" assimilates every distinction of sex, color, and rank, "The streets of New Panama presented a perpetual carnival ... for there, were assembled the people of nearly every civilized state on earth, each in his national garb, and bringing with him national peculiarities" (p. 12). In keeping with this Burkean carnival, Parker offers a fundamentally different representation of technology itself. Everything in the tale, as Gibson points out, is painted on "the grandest scale possible" (Gibson, 2019, p. 33), and the most sublime machines of all are the "vast arks" or "mighty vessels which, towards the close of the twenty-first century, bade defiance to the storms and the billows of the ocean" (Parker, 1835, p. 8). These steamships are so enormous that, when "one of those great vessels" washes up on what is now the shore of Carlos' mountain-top retreat, it takes from morning through the following day to explore "the huge edifice" (pp. 47–48). These "monsters of science" (p. 8) are, above all, far more than mammoth tools of trade and communication: they are "glorious creations of human power" (p. 8). Emphasizing the human source of their power, the tale all but ignores their mechanical nature, describing them instead as "floating islands" (p. 9) or "floating cities" (p. 48), as microcosms, that is, of human society:

> The shops, the hotels, the theatres, the markets, the baths, the places of public worship, the police, the various offices of the magistrates, and others vested with the government of this microcosm; (some vessels even bore gardens of great beauty) would no longer permit one to consider the huge fabric as a ship floating on the surface of the ocean, but rather as some maritime town, against whose bulwarks and on whose piers the billows thundered in vain. (p. 10)

After "the waters of the two oceans had mingled" and "the junction was complete," the canal opens: "Then advancing in beauty and majesty, as if a creature instinct with life and conscious of her sublime position, one of the mightiest vessels that swam on the ocean, entered from the Pacific" (pp. 35–36). We should not consider these Romantic machines as ships, the tale insists, but rather as majestic creatures "instinct with life"— and in particular, as creatures instinct with the life of humanity, their creators. In humanized technology, accordingly, Parker's anti-utilitarian perspective locates the synthesis of human and natural power. Where the utilitarian sees technology and wonders at all that its power can do, the Romantic sees "what in soul" technology is and wonders at humanity's own sublime power to create, and, of course, to destroy—in this case to create and destroy mighty empires. If machines are our creatures, not just our instruments, then our presumption can turn them into titanic monsters of science.

And through presentiments, which give voice to a wisdom without reflection that the rationalists (Carlos included) dismiss as "foolish superstitions" (Parker, 1835, p. 29), common people sense the destruction to come. While the President issued his decree assenting to the commencement of the canal, "a sudden darkness ... like the lurid dusk of a total eclipse" (p. 14) fell upon the ceremony, and the sun "appeared instinct with life, I should rather say with corruption, for it resembled a moving, heaving mass, moving with the loathsome writhing motion which you may see in a knot of vipers or earth-worms" (p. 15). Next, "a gigantic condor, grasping a mangled sea-eagle in his iron talons, fell dead in the great square" (p. 15). Along with "deep sighs" and "loud unearthly wailings," "groans and hollow murmurs were heard on the sea-shore" (p. 16). As the engineers prepare to complete the canal, Carlos "heard a sound above, below, around, filling as it were the whole atmosphere, and impregnating the very earth, of so singular a description, that it is almost impossible to convey any adequate idea of it. It was not loud, but it was heavy, solemn, mournful ... [I]t seemed as if the mountains shuddered, while a hollow groan murmured through the deep valleys, and over the sleeping ocean" (p. 28). But Carlos "was too well versed in the physical science of the age, too devout a believer in the perfection of mechanics" (p. 16), his heart too "hardened by the glories of mechanics" (p. 17), to take these signs seriously. Domingo's mother Sarah tells Carlos that this is the third time she has heard the ominous sound within the last year, but "the people of the city," she says, "laugh at us when we tell them, and call our truths, which are as true as the holy saints, foolish superstitions" (p. 29). In hindsight, he exclaims, "Alas, how has my vain imagination, my pride of knowledge, my presumptuous confidence been rebuked! The haughtiness of philosophy refused to recognize in the human mind an anticipative faculty more true than the most forecasting calculations of science" (p. 17). Having heard Sarah's account of the mournful sound, Carlos concludes that it must have been "the origin of many of those vague rumors which had agitated the public mind— they were *not* groundless—what could it mean? Were there indeed more things in heaven and earth, than we dream of in our philosophy?" (p. 29). There were: unlike the mechanized future wonderland of technological achievement in "1980," which Abhishek Sarkar characterizes as secular and materialist (Sarkar, 2023, p. 49), the dystopian world of "The Junction of the Oceans" is transcendent and providential, critiquing any purely human narrative of progress by suggesting that civilization depends upon a balance, a fittedness, between sublime human power and the earth, a balance that an immanent higher power can and will enforce. Even on the vastest of scales, our machines are nothing next to the "instruments" that God wields to "rebuke the pride of his presumptuous creature" (p. 5) when in our unremitting ardor we fail to follow our "anticipative faculty," our inborn sentiments. God therefore is a "Power whose goodness ... speaks to us" both "through all nature" and "from the depths of an irresistible conviction in our own souls, whenever we are humble, whenever we are just" (p. 4).

There is no doubt that "The Junction of the Oceans" is a tale of "technological overreaching" that expresses "deep reservations about the commercial and material consequences of empire" (Gibson, 2019, pp. 33–34), but both the specific form of the technology that overreaches and the problem posed by the conclusion of the tale indicate reservations about a particular *kind* of commercial empire, not about commerce, let alone empire, itself. Although, to be sure, the vast, humanized ships have made the canal necessary and even inevitable, it is not the ships themselves that seem to have offended God. It is rather the canal, which would alter nature in order to maximize "the commercial transactions carried on to so prodigious an extent, and so rapidly, by means of these grand vessels" (Parker, 1835, p. 10), that prompts the deluge, a phenomenon through which God speaks. As Jennifer Pitts points out with respect to Burke's "providential language"—describing the Empire as "an incomprehensible dispensation of the Divine providence," for example— "such references to the mysteriousness of the Divine plan may be read as intending to demand caution and self-doubt rather than British confidence in an imperial mission" (Pitts, 2005, p. 69). In a narrative structure analogous to the time-loops of contemporary science fiction plots, and dramatically different from the anglocentric "progress narrative" that Koditschek places at the heart of the "romance of liberal imperialism" (Koditschek, 2011, p. 100), the flood effectively resets the world (this is not the first flood, after all), suggesting that the divine plan, which we must constantly struggle to decipher in nature and ourselves, involves historical cycles of fitful, doubtful progress attained through the lessons of experience.

And the latest lesson lies in the tale's "curious ending" (Gibson, 2019, p. 34). Soon after realizing that he has "survived the miserable hour of a world's desolation" (Parker, 1835, p. 41), Carlos sees that Domingo too has survived: "I raised him—the fanciful distinction of society, and the wicked one of colour, were forgotten in a moment. I saw only the man and the brother. I raised Domingo, I pressed him to my heart, and our tears mingled ... [P]erhaps the last dark and the last white man, the sole survivors of countless myriads—we stood alone and wept" (p. 45). Twenty years later, both Carlos and Domingo have become fathers, Carlos by Henrietta, and Domingo (necessarily) by the younger Filipina.[13] The telling of the tale to Alfred is precipitated by Carlos' growing sense that he is "called by the very circumstance of our miraculous preservation" (p. 51) to perform a duty, to seek out other survivors and reset humanity, just as God has reset the earth. This sense, which he reads in his own heart, is confirmed by a sign in nature: after "the day of the world's calamity," a new "magnificent constellation" appeared for the first time in the sky. Although Alfred cannot read the "characters of lambent flame," to Carlos' own eyes "as palpable as the more brilliant of its clustered stars" is the "inspiriting legend" that they spell: "IN HOC SIGNO VINCES" ["By this sign you will conquer"] (p. 54). He asks, "Shall I again scorn in my presumption those mysterious influences which from beyond nature speak to the spirit of man, because they baffle speculation, and defy philosophy?—no! I will go forth to accomplish my destiny, in the humility of one conscious of his own weakness" (p. 55). He has built a vessel and will sail it "manned by Domingo and three of his stout sons, with two of your own brave brothers," and the party will sail "with no presumptuous dependence upon the resources of intellect or the aids of science; but with a humble reliance upon that Power into whose hands I reverently commit the issue of my enterprize" (p. 53).

The problem of the ending is that Carlos, from a contemporary perspective that perhaps wishes to find an anti-imperialist, or even proto-Marxist, critique of global capitalism, seems to have learned nothing: "it may be reserved for me to reillumine the lamp of human knowledge, to rekindle the smouldering flame of true religion amongst mankind. It may be my lot to reunite the survivors of the human race under the banners of civilization and society, to organize once more those combinations which are the elements of power, the sources and safeguards of enlightened happiness" (p. 52). Gibson suggests that the ending is an uncharacteristic instance of "missionary zeal" on the part of Parker, with Carlos as

---

[13] Henrietta's surname, Albany, emphasizes the European family's whiteness alongside the "darkness" of the "negro" Domingo and the Filipina Creole's family. ("Black" does not appear in the tale as a racial descriptor.)

"an imperial proselytizer," and asks, "How can we judge whether Parker's conclusion is deliberately grandiloquent to the point of irony or whether we are expected to sympathize with Carlos's quite possibly futile quest?" (Gibson, 2019, p. 35). Whether or not we can, I think that we are expected to sympathize with Carlos's quest, precisely because it is not just possibly futile but self-proclaimedly and (like the future) emphatically uncertain: unlike the rationalist he was, he will now eschew all "presumptuous dependence upon the resources of intellect or the aids of science" and will instead try to read and follow the signs of a divinely but mysteriously ordered world, and that, for Parker, is all that anyone can hope to do.

So Gibson is right that the tale is "Parker's intervention against those who glorify international commerce. It sends up the notion that the advancement of commerce is identical with the advancement of civilization" (Gibson, 2019, p. 36). More specifically, though, it is an intervention against two utilitarian perspectives upon that equation. First, it is a critique of those who trust their own systematic reasoning and believe that if the advancement of commerce is identical with the advancement of civilization, then the maximization of the former will maximize the latter and is therefore justifiable at any expense. And second, it is thoroughly skeptical about the existence of civilization separate or apart from its local manifestations and modes of development in particular circumstances as it relocates from one-quarter of the globe to another. Recall that when civilization comes to New Panama, it assembles "the people of nearly every civilized state on earth, each in his national garb, and bringing with him national peculiarities." For Parker, then, Carlos has learned something crucial, and it is the Romantic lesson embedded in the thoroughly Burkean language of the passage with which the preceding paragraph began: Carlos will not diffuse knowledge, he will reillumine; Carlos will not convert, he will rekindle; he will not assimilate, absorb, or annex the survivors of the human race, he will reunite them. He will organize "once more" the combinations of humanized technology and commerce, which *are* in fact the elements of power and the sources of human happiness so long as they proceed from a spirit of humility and so long as they fit with the power of nature, thereby fulfilling "the will of heaven" (p. 46). He needs to listen to the earth, and to history, and to experience, in order "to discover, to console, to reanimate with hope those fragments of the human race which survived ... to point out for future generations the track to those vast realms which reflection assures me must have been created by—THE JUNCTION OF THE OCEANS" (p. 55). Thus ends the tale, in 2074, with a vessel bearing seven cosmopolitans of sentiment, four "dark" and three "white," on an uncertain voyage to new vast realms in order to resume, this time not in a spirit of "boundless utility" but "in the humility of one conscious of his own weakness," the empire of the middle classes.

REFERENCES

Baynham, Walter. (1892). *The Glasgow Stage*. Glasgow: Robert Forrester.

Boase, Frederic. (1965). *Modern English Biography: Containing Many Thousand Concise Memoirs of Persons Who Have Died Between the Years 1851–1900*. 2nd ed., vol. 6. London: Frank Cass.

Buckland, C.E. (1906). *Dictionary of Indian Biography*. London: Swan Sonnenschein.

"Calcutta Pococurante Society." (1833). *The Calcutta Quarterly Magazine and Review*, 3, [1–60].

Campbell, Donald. (1996). *Playing for Scotland: A History of the Scottish Stage 1715–1965*. Edinburgh: Mercat Press.

Chanda, Mrinal Kanti. (1987). *History of the English Press in Bengal 1780 to 1857*. Calcutta: K.P. Bagchi.

Chaudhuri, Rosinka. (2002). *Gentlemen Poets in Colonial Bengal: Emergent Nationalism and the Orientalist Project*. Calcutta: Seagull Books.

———. (2018). "'On the Colonization of India' (1829): Public Meetings, Debates and Later Disputes." *The Indian Economic and Social History Review*, 55(4), 463–89.

Corfield, Wilmot. (1913). "Henry Meredith Parker." *Notes and Queries* 11(7)(179), 436.

Cox, Jeffrey N., ed. (1999). *Slavery, Abolition and Emancipation: Writings in the British Romantic Period*. Vol. 5, *Drama*. London: Pickering & Chatto.

"Dacca Marriages 1822." (1822). British Library India Office Records N/1/12 f.197.

Darwin, Erasmus. (1791). *The Botanic Garden*. London: J. Johnson.

Doss, Ramchunder. (1844). *A General Register of the Hon'ble East India Company's Civil Servants of the Bengal Establishment from 1790 to 1842*. Calcutta: Baptist Mission Press.

Dover, Cedric. (1937). *Half-Caste*. London: Martin Secker and Warburg.

Eden, Emily. (1872). *Letters from India*. Edited by Eleanor Eden, vol. 1. London: Richard Bentley and Son.

Ehrlich, Joshua. (2016). "The Calcutta Pococurante Society: Public and Private in India's Age of Reform." *The Public Domain Review*. https://publicdom ainreview.org/essay/the-calcutta-pococurante-society-public-and-private-in-indias-age-of-reform/.

"Free Press Dinner at the Town Hall." (1838). *The Calcutta Monthly Journal*, 39, 82–92.

Gibson, Mary Ellis. (2011). *Indian Angles: English Verse in Colonial India from Jones to Tagore*. Athens, GA: Ohio University Press.

———, ed. (2019). *Science Fiction in Colonial India, 1835–1905: Five Tales of Speculation, Resistance and Rebellion*. London: Anthem Press.

Grant, Colesworthey. ([1850?]). "H.M. Parker, Esq." *Lithographic Sketches of the Public Characters of Calcutta. 1838–1850*. Calcutta: W. Thacker and Co.

Highfill, Philip H., Jr. (1982). *A Biographical Dictionary of Actors, Actresses, Musicians, Dancers, Managers & Other Stage Personnel in London, 1600–1800. Volume 8, Hough to Keyse*. Carbondale: Southern Illinois University Press.

"H.M. Parker, Esq. Bengal Civil Service." (1840). *India Review and Journal of Foreign Science and the Arts*, 4, 589–97.

"The Humble Petition of Henry Meredith Parker." (1814). British Library India Office Records J1/29/14-22 1814.

Kling, Blair B. (1976). *Partner in Empire: Dwarkanath Tagore and the Age of Enterprise in Eastern India*. Berkeley: University of California Press.

Koditschek, Theodore. (2011). *Liberalism, Imperialism, and the Historical Imagination: Nineteenth-Century Visions of a Greater Britain*. Cambridge: Cambridge University Press.

Makepeace, Margaret. (2021). "East India Company Appointments by the Prince Regent—(1) Henry Meredith Parker." *British Library Untold Lives Blog*. https://blogs.bl.uk/untoldlives/2021/08/east-india-company-appointments-by-the-prince-regent-1-henry-meredith-parker.html.

"Marriages within the Chaplaincy Station or District of Calcutta in the Archdeaconry and Diocese of Calcutta." (1832). British Library India Office Records N/1/33 f.205.

"Metcalfe Meeting." (1838). *The Calcutta Monthly Journal*, 39, 94.

Mittra, Kissory Chand. (1870). *Memoir of Dwarkanath Tagore*. Calcutta: Thacker, Spink and Co.

"Mr. Buckingham's Petition. To the Honourable the Commons of the United Kingdom of Great Britain and Ireland, in Parliament Assembled." (1826). London: Mills, Jowett and Mills.

Muir, Rory. (2019). *Gentlemen of Uncertain Fortune: How Younger Sons Made Their Way in Jane Austen's England*. New Haven: Yale University Press.

Ní Fhlathúin, Máire. (2015). *British India and Victorian Literary Culture*. Edinburgh: Edinburgh University Press.

"On the Colonization of India by Europeans" [by "S.J."]. (1978). In Gautam Chattopadhyay, ed., *Bengal: Early Nineteenth Century (Selected Documents)*. Calcutta: Research India Publications, pp 32–35.

"On the Proposed Communications Between the Atlantic and Pacific Oceans." (1824). *The Oriental Herald and Colonial Review*, 1(3), 407–17.

Parker, Henry Meredith. (1827). *The Draught of Immortality, and Other Poems: With Cromwell, a Dramatic Sketch.* London: J.M. Richardson.

———. (1835). "The Junction of the Oceans. [*A Tale of the Year 2074.*]." In David Lester Richardson, ed., *The Bengal Annual. A Literary Keepsake for MDCCCXXXV.* Calcutta: Samuel Smith, pp. 1–55.

———. (1851a). *Bole Ponjis. Containing The Tale of the Buccaneer; A Bottle of Red Ink; The Decline and Fall of Ghosts; and Other Ingredients.* Vol. 1. London: W. Thacker.

———. (1851b). *Bole Ponjis. Containing The Tale of the Buccaneer; A Bottle of Red Ink; The Decline and Fall of Ghosts; and Other Ingredients.* Vol. 2. London: W. Thacker.

———. (1858a). *Caste and Conversion: Being No. 3 of Short Sermons on Indian Texts, Concerning the Empire of the Middle Classes.* London: W. Thacker.

———. (1858b). *The Empire of the Middle Classes. Being Nos. 1 and 2 of Short Sermons on Indian Texts.* London: W. Thacker.

———. (1858c). *A Plan for the Home Government of India. With Provisions Calculated to Prevent or Limit the Dangers and Evils of Patronage.* London: W. Thacker.

Parkes, Fanny. (2001). *Wanderings of a Pilgrim in Search of the Picturesque.* Edited by Indira Ghose and Sara Mills. Manchester: Manchester University Press.

Pitts, Jennifer. (2005). *A Turn to Empire: The Rise of Imperial Liberalism in Britain and France.* Princeton: Princeton University Press.

Ray, Indrajit. (2011). *Bengal Industries and the British Industrial Revolution (1757–1857).* London: Routledge.

Reports from Committees: Colonization and Settlement (India). (1859). Parliamentary Papers Session 3 February–19 April 1859.

Sarkar, Abhishek. (2023). "The Celebration of Technoscience in '1980,' a Futuristic Science Fiction Published from Calcutta in 1835." *ANQ: A Quarterly Journal of Short Articles, Notes and Reviews,* 36(1), 45–50.

Serajuddin, A.M. (1978). "The Salt Monopoly of the East India Company's Government in Bengal." *Journal of the Economic and Social History of the Orient,* 21(3), 304–322.

Stocqueler, J.H. (1873). *The Memoirs of a Journalist.* Bombay: The Office of the Times of India.

White, Daniel E. (2013). *From Little London to Little Bengal: Religion, Print, and Modernity, 1793–1835.* Baltimore: Johns Hopkins University Press.

———. (2016). "'Zig Zag Sublimity': John Grant, the Tank School of Poetry, and the *India Gazette* (1822–1829)." In Rosinka Chaudhuri, ed., *A History of Indian Poetry in English.* Cambridge: Cambridge University Press, pp. 147–61.

INDEX

GPSR Compliance

The European Union's (EU) General Product Safety Regulation (GPSR) is a set of rules that requires consumer products to be safe and our obligations to ensure this.

If you have any concerns about our products, you can contact us on ProductSafety@springernature.com

In case Publisher is established outside the EU, the EU authorized representative is:

Springer Nature Customer Service Center GmbH
Europaplatz 3
69115 Heidelberg, Germany

The manufacturer's authorised representative in the EU is Springer
Nature Customer Service Centre GmbH, Europaplatz 3, 69115 Heidelberg,
Germany. If you have any concerns regarding our products, please
contact ProductSafety@springernature.com

Printed and bound by CPI Group (UK) Ltd, Croydon, CR0 4YY

29/04/2026

02099545-0004